Handling Qualities Technical Meeting 2023

Huntsville, Alabama, USA
7 - 8 February 2023

ISBN: 978-1-7138-8785-0

Printed from e-media with permission by:

Curran Associates, Inc.
57 Morehouse Lane
Red Hook, NY 12571

CURRAN ASSOCIATES INC. proceedings.com

Some format issues inherent in the e-media version may also appear in this print version.

Vertical Flight Society
2700 Prosperity Ave, Suite 275
Fairfax, VA 22031
USA

Phone: (703) 684-6777
Fax: (703) 739-9279

www.vtol.org

Additional copies of this publication are available from:

Curran Associates, Inc.
57 Morehouse Lane
Red Hook, NY 12571 USA
Phone: 845-758-0400
Fax: 845-758-2633
Email: curran@proceedings.com
Web: www.proceedings.com

TABLE OF CONTENTS

Vertical Flight Society
REDSTONE CHAPTER

VFS Handling Qualities Technical Meeting Agenda
February 7-8, 2023
Radiance Technologies, Huntsville, AL

	Tuesday, February 7, 2023
0700-1200	**Attendee Registration**
0700-1200	**Presentation Material drop-off for Authors**
0800-0815	**Welcome by Chapter President –** Dr. Bill Lewis, TTTT
	Meeting Objectives by General Chair – Mr. Eric Grigorian, GTRI
0815 - 0900	**Opening Keynote – Carl R. Ott (XP), Stephen T. Kubik** **UH-60Mx Fly-by-Wire Degraded Visual Environment Handling Qualities Testing**
	Session I – Test and Evaluation
	Session Chair – Kit Borden, US Army CCDC-AvMC
0900-0910	Session Chair Remarks
0910-0930	*Development of the RASCAL X Flying Test Bed* (*Presentation Only*), Brian T. Fujizawa, MAJ Wesley R. Ogden, U.S. Army DEVCOM
0930-0955	*Investigation of Pilot Inceptor Metrics as Correlates to Qualitative Workload,* Ryan Paul, Matthew Rhinehart, Kenneth Cahill, Zachary Smith, Oklahoma State University / NAVAIR
0955-1010	**Break**
1010-1035	*U.S. Army eVTOL Qualitative Evaluation*, MAJ W. Ogden, U.S. Army CCDC-AvMC
1035-1100	*eVTOL Flight Test Council Handling Qualities Means of Compliance Advisory Committee* (*Presentation Only*), Anthony Mumford, Martin Shubert, David Klyde, Overair, Inc
1100-1125	*Turbulence Cancelling Technology for Reliable and Predictable AAM Operation (Presentation Only)*, DI Andras Galffy, DI Ines Burgstaller, Oliver Breiteneder, Turbulence Solutions
1125-1210	Questions and Roundtable Discussion
1210-1300	**Lunch**
	Session II – Flight Controls
	Session Chair – Dr. Jeff Lusardi, ST, US Army CCDC-AvMC
1300-1325	Session Chair Remarks: *Handling Qualities Requirements in Design*
1325-1350	*On the Characterization of Nonlinear Frequency Interactions of the Pilot Vehicle System (Rotorcraft): A Preliminary Study*, Robert Alstrom PhD, Ryan Paul PhD, Georgia Tech / Oklahoma State
1350-1405	**Break**
1405-1430	*Rotorcraft Vertical Rate Command's Impact on Pilot Workload and Handling Qualities and its Implementation-Practicality (Presentation Only)*, Ian Willebeek-LeMair, Matt Rhinehart, NAWCAD
1430-1455	*MIL-DTL-32726 Status (Formerly ADS-33E) (Presentation Only)*, F. Scott Smith, US Army CCDC AvMC
1455-1540	Questions and Roundtable Discussion
1540-1730	**Social**

Vertical Flight Society
REDSTONE CHAPTER

VFS Handling Qualities Technical Meeting Agenda
February 7-8, 2023
Radiance Technologies, Huntsville, AL

Wednesday, February 8, 2023

0700-1200	**Attendee Registration**
0700-1200	**Presentation Material drop-off for Authors**
0800-0815	**Day 2 Welcome by General Chair –** Mr. Eric Grigorian
	Session III – Human Factors
	Session Chair – Bill Lewis, Tennessee Technical Test Team
0815-0825	Session Chair Remarks
0825-0850	*Human Engineering Considerations in Automated Flight (Presentation Only),* Bob Copeland, U.S. Army AvMC
0850-0915	*Predicting Pilot Performance, Workload and Handling Qualities using SCOPE*, Edward Bachelder, Bimal Aponso, Tom Berger San Jose State University Research Foundation/FAA/U.S. Army CCDC-AvMC
0915-940	*Handling Qualities Evaluations of Active Inceptors with Varying Force-Feel Characteristics*, Michael Jones, David Klyde, Dakota Musso, Cody Fegely, Luke Gillett, Seung Baik Systems Technology, Inc / Bell Textron Inc.
0940-0955	**Break**
0955-1020	*Inter- and Intra- Pilot Rating Variability during Dynamic Interface Simul*ation, Jeremy Scroggs, Jonathan Dronfield, Zachary Smith, NAVAIR
1020-1055	*Reliability, Risk, and Safety Considerations for Architectures that include Artificial Intelligence, including Verification and Validation of AI Algorithms (Presentation Only),* Brenda Menees, US Army CCDC, AvMC
1055-1120	*Theoretical Basis of Adverse Pilot Coupling (Presentation Only)*, Edward Bachelder, San Jose State University Research Foundation,
1055-1140	Questions and Roundtable Discussion
1145-1230	**Lunch**
1230-1315	**Day 2 Keynote – Rotorcraft-in-flight Laboratory Overview, MAJ Wes Ogden U.S. Army CCDC-AvMC**
1315 - 1345	**Closing**– President, Redstone Chapter

RADIANCE TECHNOLOGIES
100% EMPLOYEE-OWNED

Hosted by ***Radiance Technologies*** *at 310 Bob Heath Dr., Huntsville, AL 35806*

Investigation of Pilot Inceptor Metrics as Correlates to Qualitative Workload

Dr. Ryan Paul
Assistant Professor
Oklahoma State University
Stillwater, OK, USA

Matthew Rhinehart
Aerospace Engineer
NAVAIR
Patuxent River, MD, USA

Kenneth Cahill
Aerospace Engineer
NAVAIR
Patuxent River, MD, USA

Zachary F. Smith
Aerospace Engineer
NAVAIR
Patuxent River, MD, USA

ABSTRACT

Advances in control system and augmentation technologies in the recent decades have allowed major advances in both the safety and mission effectiveness of military aircraft. These technologies have also resulted in an increase in the complexity of the development and certification process, partially due to the myriad of new "knobs" that can be turned. At the same time, the primary means of Handling Qualities evaluation remains subjective, via pilot assigned ratings, unchanged since Cooper & Harper published their scale in 1967. This effort sought to collect a large quantity of fundamental piloted simulation data, primarily to establish if correlation exists between qualitative rating scales and previously developed inceptor workload metrics. Secondarily, different types of classical Handling Qualities degraders were used to determine if different inceptor metrics were better able to identify them. Results indicate that most pilot subjects tested show at least some correlation between qualitative ratings and quantitative inceptor metrics, however few showed strong enough correlation to consider usurping the role of qualitative scales. Further, it showed that to achieve any meaningful correlation, treating pilots as individuals is necessary, but not sufficient, as a few pilots demonstrated poor or no correlation.

INTRODUCTION

Advances from near-peer adversaries and the drive to ever-improve warfighter safety serve to apply pressure to increase the capability of military aircraft. One major advance to answer this call has been the development of cutting-edge fly-by-wire control systems. These systems deliver unheralded flexibility and capability; however, it has resulted in significant increases in the time required to optimize the handling qualities. This is partly due to the bewildering number of possible tunable parameters that can be applied across ever-broadening flight envelopes. At the same time, the drive to develop and field aircraft faster and more cheaply are in direct opposition to the desire for additional capability. Accurately discerning the workload required by a pilot for completing a given mission task or mission task element is important for aircraft certification, development of good handling qualities, and assuring mission effectiveness. A robust and objective workload metric would not only offer quick verification of handling qualities, it could also serve to guide the control designer to which aspects of the design to tune in the development of capability, thus increasing efficiency.

Historically, pilot workload is given based on qualitative scales, such as the Cooper-Harper Handling Qualities Rating (CHR) scale (Ref. 1). The CHR Scale is a 10-point ordinal scale, and through a series of questions, guides the pilot to a rating. The CHR Scale is a composite of task performance and self-assessed workload. While task performance can be objectively measured, self-assessed pilot workload is subjective. Measurement errors stem from factors such as inter- and intra-pilot variation, non-linearities in the CHR Scale, and other issues as discussed further in Ref. 2 and Ref. 3. Other intangible challenges arise when trying to measure workload objectively such as learning effects or piloting strategy changes. In addition, differences between simulation and flight, such as modeling issues, cueing limitations, environmental factors, adherence to aircraft limitations, and self-preservation, pose an increased challenge to objectively

Presented at the VFS Handling Qualities Technical Meeting, Huntsville, Alabama, USA Feb. 7-8 2023.

measuring workload in pilot-in-the-loop simulators; which are often used during the aircraft development process. These subjective variations often limit the utility of the CHR Scale for the desired purpose, leading to the question of "how to measure workload objectively?".

Objective measurement of workload has long been of interest both in the field of handling qualities and other disciplines producing concepts spanning physiological, neurological, aircraft state, inceptor activities, and other measures as shown in Refs. 3, 4, and 5. This study continues the investigation of using pilot inceptor workload metrics as an analog to subjective workload due to ease of collection from both flight and simulation. Many inceptor measures have been proposed and evaluated to varying degrees of rigor, however none have found their way into regular use by the handling qualities community. While the general inertia of the status quo may partially be to blame, most of the proposed inceptor activity measures are not suitable for broad application, and therefore are not widely adopted. With the exception of the aggressiveness metric in Ref. 5, most metrics focus on correlation of a single axis with little discussion on application in multi-axis tasks. In addition, limited information is available in the literature regarding application of metrics to time varying tasks. Ref. 6 applies the inceptor based workload metrics that could be found (Ref. 4 and 5, and Ref. 7 through Ref. 12) to a given set of data in an attempt to determine which ones offer the greatest advantage. While initial findings were inconclusive, it was thought this was due to the task against which the metrics were run and the size of the candidate pool among other reasons. Therefore, data collected specifically for analysis against these metrics was needed.

For a relatively complicated task such as the Aeronautical Design Standard 33 (ADS-33) Precision Hover Mission Task Element (MTE) (Ref. 13), a viable metric must objectively determine workload during the 45-degree inbound translation, hover capture, and position hold phases. It is not immediately clear if the composite handling quality rating (HQR) for the task should be based on the average or peak value. The United States Naval Test Pilot School (USNTPS) does not explicitly instruct students to consider peak or average workload over the total task. This leaves an open question, which could contribute to both inter- and intra- pilot variability. Finally, a functional workload metric must be able to determine workload change stemming from different factors such as control tuning parameters, environmental effects, and cueing environment. A relevant workload metric should be scalable to any number of axes, not require knowledge of the explicitly observed error by the pilot (pursuit vs. compensatory), account for the time varying nature of some tasks, and adequately capture different drivers of workload. The study laid out within this report uses data collected on a relatively large candidate pool that had pilots fly tasks that were derived from selected MTEs with known degraders to drive up workload. Collecting and analyzing a large pool of data is expected to reveal the capability of the PIW metrics to stand-in as an analog to workload.

METHODOLOGY

In an effort to modernize the approach to measure pilot workload, this research followed a novel path for analyzing pilot inceptor data to produce results that correlate with perceived workload. This work was predicated on the hypothesis that either a singular metric or population of metrics exist that can be computed from recorded data whose values correlate with qualitative pilot workload. Following an initial examination using historical data from a previous research endeavor (Ref. 6), the correlation between qualitative workload and computed metrics was weaker than desired and attributed in part to being restricted to only HQRs as the measure of qualitative workload. The project then shifted focus and emphasized developing and executing a clinical set of tasks in a simulator to build a more comprehensive database of qualitative pilot assessments using three different techniques for capturing pilot workload. The final product was a database of 12 subject matter expert pilots performing up to 5 tasks in a pilot-in-the-loop simulator to provide more insight into the relationship between the task, inceptor motions, and perceived workload. Each MTE was performed multiple times with varying aircraft configurations to elicit higher workload in manners known to degrade aircraft handling characteristics as documented in ADS-33. The research goal was to assess the appropriateness of various pilot inceptor workload (PIW) metrics and learn how to apply them in a test environment.

The simulation tasks flown were primarily designed to collect a broad set of data to conduct research into an objective measurement of pilot workload. The five tasks selected emphasized differences between single vs. multiple axis tasks, compensatory vs. pursuit tasks, pseudo-time variant vs. time varying tasks, and fine vs. large control input tasks. Each task was first flown using a nominal air vehicle configuration, and subsequently repeated, each time adjusting the underlying model to exhibit classical degraders of handling qualities, per ADS-33 specifications.

The Simulation Environment

The host lab for this experiment was the single seat fixed base simulator at the USNTPS. The cockpit was generic and equipped with inceptors capable of supporting fixed or rotary wing simulations. A control loader provided functionality to adjust mechanical characteristics of the inceptors. The typical center stick and collective helicopter controls were used for this evaluation. A flat panel computer monitor in the cockpit provided representational flight instruments, while a large flat projection, approximately 8 feet square, presented the pilot with an out the window visual scene generated with a single ceiling mounted projector. Figure 1 shows the cockpit arrangement and Figure 2 exhibits the typical setup during data collection.

Figure 1. Simulator Pilot Seat, Controls, and Primary Flight Display.

Figure 2: Overall Simulation Configuration.

The flight dynamics model for this research was of state space form and loosely based on the OH-6. Several underlying assumptions limiting the applicability of the model are documented here. All of these assumptions can be removed by increasing the complexity of the model, but at the cost of loss of insight and ease of customization. The significant assumptions for the model are as follows:

- The model is for a fixed operating point with constant coefficients. Stability derivatives are used to define the response characteristics.
- The model is valid only for hover and low-speed flight.
- Some stability derivatives have been neglected. The most significant of these are the static stability derivatives Mw and Nv, along with the speed-dependent derivatives Zu, Mu, and Lv. The main consequence of this simplification is that the free response dynamics are not in the form of the familiar hovering cubic that is typical for three degree of freedom models (Ref. 1).
- All inter-axis coupling derivatives are set to zero. Placeholders are identified for pitch-to-roll and roll-to-pitch coupling due to aerodynamics (Mp and Lq) or control mixing (M_A and L_B). These derivatives should be set to achieve the desired amount and type of coupling desired, as discussed later in this paper.
- Control derivatives are considered variables to be set in simulation to provide best response for each set of dynamics. Alternatively, multipliers can be included in the simulation environment allowing for adjustment of control sensitivity without modifying the values of the derivatives.

A first-order state-space representation of a hovering helicopter with fixed aerodynamic parameters comprises the basic math model. Derivatives that define oscillatory low-frequency (phugoid) dynamics are not included, though it is possible to increase model complexity and add those derivatives. Capability for pitch-due-to-roll and roll-due-to-pitch coupling, resulting from both dynamics and control coupling, is included in the model. In addition, an Attitude Command model was generated by adding angular attitude derivatives to the equations of motion. The resulting generalized state-space matrices for an $\dot{x} = Ax + Bu$ state transition form are as given below, included as Figure 3:

$$A = \begin{bmatrix} X_u & 0 & 0 & -g & 0 & 0 & 0 & 0 \\ 0 & Z_w & U_0 & 0 & 0 & 0 & 0 & 0 \\ 0 & 0 & M_q & M_\theta & 0 & M_p & 0 & 0 \\ 0 & 0 & 1 & 0 & 0 & 0 & 0 & 0 \\ 0 & 0 & 0 & 0 & Y_v & 0 & -U_0 & g \\ 0 & 0 & L_q & 0 & 0 & L_p & 0 & L_\phi \\ 0 & 0 & 0 & 0 & 0 & 0 & N_r & 0 \\ 0 & 0 & 0 & 0 & 0 & 1 & 0 & 0 \end{bmatrix} \text{ and } B = \begin{bmatrix} 0 & 0 & 0 & 0 \\ 0 & Z_{\delta_C} & 0 & 0 \\ M_{\delta_B} & 0 & M_{\delta_A} & 0 \\ 0 & 0 & 0 & 0 \\ 0 & 0 & 0 & 0 \\ L_{\delta_B} & 0 & L_{\delta_A} & 0 \\ 0 & 0 & 0 & N_{\delta_P} \\ 0 & 0 & 0 & 0 \end{bmatrix}$$

$$X = \begin{bmatrix} u \\ w \\ q \\ \theta \\ v \\ p \\ r \\ \phi \end{bmatrix} \quad \text{and} \quad u = \begin{bmatrix} \delta_B \\ \delta_C \\ \delta_A \\ \delta_P \end{bmatrix}$$

Figure 3: State Space Matrices.

Variations in vehicle response (bandwidth) are made by changing the values of the dynamic derivatives. Control derivatives in the B matrix were adjusted in simulation to provide best control/response based on pilot comments during model development. Simplifications to the baseline A and B matrices were made to establish a canonical model with fully decoupled state and control matrices. All modal dynamics were along the real axis in an s-plane representation. Artifacts such as rotor delay and higher order lag states were excluded, however an unstable long period pitch and roll mode was incorporated to reflect the natural hovering instability of a helicopter. Implementation of the unstable long-period dynamics in pitch and roll was done by introducing Mu and Lv terms to the A matrix.

As a baseline, the resulting model was tuned to fall near/on the level 1 boundary of ADS-33 criteria, as shown for short-term response criteria and pitch/roll oscillation limits in Figure 4. To accommodate analytical handling qualities bandwidth determination, a delay term was added to the baseline model to drive the phase curve to roll off beyond -180 degrees at high frequencies. The delay value selected was 1/60 seconds, which was the approximate update rate of the visual projector. No delays were added to the Simulink model, which executes in real-time during testing, as the equivalent visual delay was present while running the experiment. There was no ambient wind present in the simulator environment.

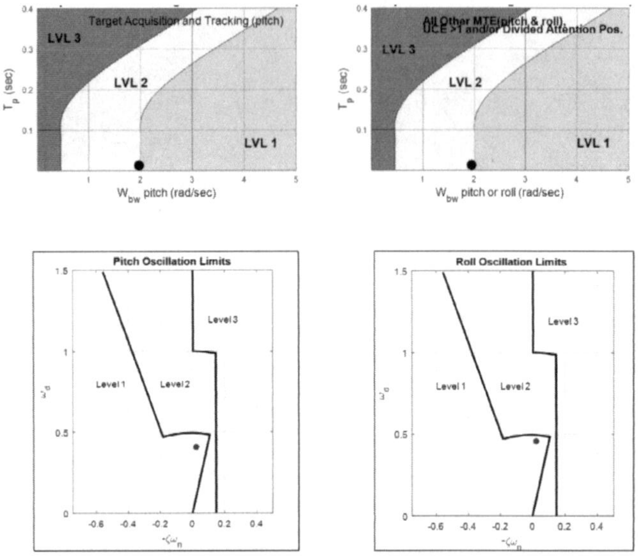

Figure 4: Baseline Model Characteristics.

To drive workload, the baseline Level-1 model was degraded consistent with deficiencies documented in ADS-33. Modifications to the plant model stability, short term response bandwidth, cross coupling, and response type were enacted simply by modifying terms in the *A* and *B* matrices defined previously. The nonlinear model modifications, an introduction of phase delay and a rate limit, were enacted directly in the underlying Simulink model. Table 1 summarizes the various modifications applied to the model and explored during the testing.

Table 1: Model Modifications.

Modification	Description
Reduced Stability	Degraded overall model to level 2 HQ
Phase Delay	Introduced 0.2s time delay
Response BW	Degraded BW to level 2 HQ
Cross Coupling	Increased coupling to level 2 HQ
Rate Limit	Introduced a 1"/s rate limit on the cyclic
Stick Force	Increased stick force gradient to level 1 limit
Response Type	Enable attitude response type

Task Descriptions: Sum of Sines - Pitch, Roll, Pitch& Roll

The Sum of Sines (SoS) task was driven by an automated command signal generated by the simulation environment. From steady, wings level flight, pilots were instructed to aggressively track the displayed signal and attempt to keep errors within the specified tolerances specified in Table 2, as adopted from Ref. 13. This maneuver did not require a test course but rather a visual cue with the desired and adequate performance criteria displayed to the pilot to enable real-time compensatory tracking. The length for scoring time was 60 seconds. There was a 10 second ramp in/ramp out period on each end of the 60 second scoring window. Figure 5 is representative of the cues provided during the sum of sines task. The pitch indication is the green dot at the center of the image. Desired for pitch is to keep the green dot within the inner circle while adequate is to keep it within the magenta circle. The roll indication is the green horizontal line. Desired for roll is to keep the green line within the inner most wedge shape while adequate is to keep it within the magenta wedge shape. This task was driven by an automated command signal generated by the simulation environment and followed a randomized sum of sines command.

Table 2: Desired and Adequate Tolerances.

	Desired	Adequate
Pitch: At least X% of the scoring time	50%	75%
within pitch attitude error tolerance:	±1°	±2°
Roll: At least X% of the scoring time	50%	75%
within roll attitude error tolerance:	±5°	±10°
PIO Considerations	No PIO tendencies	No divergent PIO tendencies
Inter-axis coupling shall not be:	Undesirable	Objectionable

Figure 5: Sum of Sines Cueing.

Task Descriptions: Point to Point Reposition

Pilots were initialized in a 20-foot hover over the runway threshold on centerline and holding the runway cardinal heading. They were then instructed to maneuver the aircraft to arrive in a hover at an altitude of approximately 20 feet, 1000 feet down the runway on a reciprocal heading and on centerline. At some point during the maneuver, the pilot was to climb to approximately 100 feet. The pilot would call "mark" when they departed the initial hover location and "mark" again when they arrived at the second hover location to the pilot's satisfaction. The test course utilized standard ICAO runway markings as shown in Figure 6. As this task was intended to allow the pilot flexibility in how it is completed (and to what accuracy), no absolute performance standards were prescribed. The description of the maneuver specified beginning and ending positions and headings, as well as an intermediate altitude, but the pilot was asked to follow these directions to their own level of acceptable tolerances. An order of operations was not specified (e.g. the pilot can choose to effect the heading change or climb to 100 feet at any time during the maneuver). The elapsed time from "mark" to "mark" was, recorded however no time requirements were dictated and there should be no assumption that a faster time was better. The pilot was expected to use their nominal control strategy and let the natural level of workload fallout.

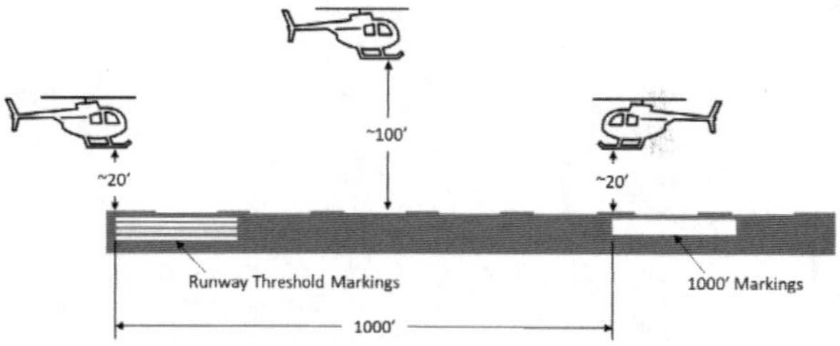

Figure 6: Runway Markings.

Task Descriptions: Precision Hover

The pilot initiated the maneuver at a ground speed of between 6 and 10 knots, at an altitude less than 20 feet. The target hover point was oriented approximately 45 degrees relative to the heading of the rotorcraft. The target hover point was a repeatable, ground-referenced point from which position deviations were assessed. The test course for this maneuver is depicted in Figure 7. Note that differences between the eye-point longitudinal position relative to the target longitudinal position will effectively change the tightness of the lateral and vertical tolerances (closer equals larger tolerances). Pilots accomplished the transition to hover in one smooth maneuver. It was not acceptable to accomplish most of the deceleration well before the hover point and then to creep up to the final position.

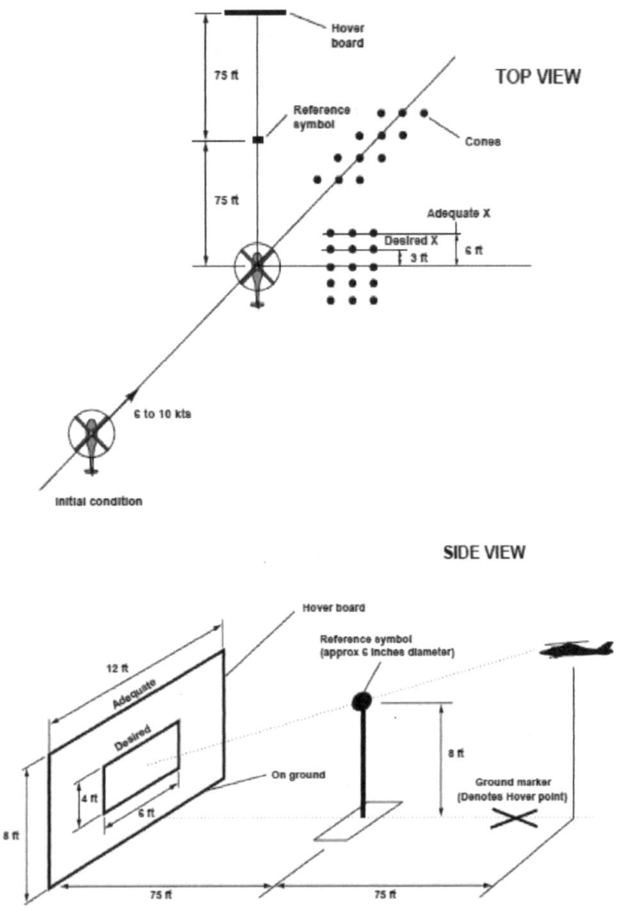

Figure 7: Precision Hover Course.

Subject Matter Expert Characterization

The test group was a diverse body of pilots from civilian and military training. Civilian pilots were all at least private pilot rated, three of which held instrument ratings with one holding a commercial flight instructor rating. Two of the civilians have received formal training from a military test pilot school. All of the pilots in the civilian group primarily had single engine piston time with an average flight experience of approximately 650 hours. Military pilots were predominately recent graduates of USNTPS in their first test tour with fleet experience in fixed and rotary wing aircraft that spanned F-18E/F/G, AV-8B, P-3C, P-8A, E-2D, and UH-1Y. Exceptions include one subject who was a retired Marine Corps CH-53E fleet aviator yet another was a current UH-60L Army National Guard pilot. The average flight experience among military pilots was approximately 1600 flight hours. The test group is outlined in Table 3.

Table 3: Pilot Backgrounds.

Evaluator	Type	Primary Platform	Total Time (hrs.)	Active Pilot	Highest Aviation Credential
1001	GA	Van's RV-4	473	Yes	PPL + IR
1002	GA	PA-28	305	Yes	PPL
1003	GA/Mil	C-172/ M-20C/ UH-60L	550	Yes	Fleet Army Co-pilot
1004	Mil	P-3/P-8	1180	Yes	Test Pilot
1005	Mil	E-2	2105	Yes	Test Pilot
1006	Mil	UH-1Y	1660	Yes	Test Pilot
1007	Mil	F-18	1000	Yes	Test Pilot
1008	Mil	CH-53E	1900	No	Fleet Marine Aircraft Commander
1009	GA	C-150	1080	Yes	PPL+IR
1010	GA	C-172	60	No	PPL
1011	Mil	AV-8B	1805	Yes	Test Pilot
1012	Mil	P-3/P-8	1652	Yes	Test Pilot

Workload Rating Scales

Classically, handling qualities and more specifically workload, have been evaluated in qualitative ways. Rating scales such as Cooper-Harper, Bedford, and NASA task load index (TLX) can be used to guide pilots to a subjective rating to characterize the goodness of the system to accomplish a task. For this evaluation, pilots were asked to provide qualitative responses in accordance with selected rating scales as outlined below. Pilots were not asked to provide Bedford workload ratings (BWR) because it requires the definition of a secondary task with which to judge spare capacity. There was a concern that the addition of a secondary task could color the results by fundamentally modifying the primary task and inadvertently correlate ratings with the Plain Old Workload Scale (POWS). In other words, while providing a secondary task as a measure of primary task workload is certainly a valid approach (as with Bedford), there was concern that the secondary task would inappropriately affect the quantitative measures of workload by altering the pilots' inceptor activity characteristics. For each task and model modifier type, the pilot was asked to provide qualitative workload according to three different rating schemes: POWS, NASA TLX, and CHR.

Workload Rating Scales: Plain Old Workload Scale (POWS)

POWS, which was created for this effort, was a contrived 10-point linear scale with terminal points anchored by zero effort on the low side and maximum effort on the high side. Task performance was not explicitly considered in this rating although it was expected that the subject intrinsically considered their performance in the selection of their POWS rating. Subjective handling qualities or workload scales try to address the inherent issues in qualitatively assigned ratings by applying some form of structure to anchor results from different tests and pilots. For this effort, the possible benefit of level setting all subjects is acknowledged but its necessity is not pre-supposed. Removing the requirement that pilots' ratings should be directly comparable allows the application of a minimally structured, ordinal scale. In the case of POWS, the pilot was asked to rate their workload on a scale of 1 (being negligible workload) to 10 (cannot work any harder). It is left up to the pilot to determine the workload make-up from various sources such as physical, mental, etc. but it does not matter as long as the pilot provided a number between 1 and 10. The pilot may give half ratings, such at 4.5, if deemed necessary.

Workload Rating Scales: NASA Task-Load Index (TLX)

TLX is a NASA-developed approach (described in Ref. [1]) that first weighs a series of factors according to the pilot's perception of importance as they related to accomplishing the task. The pilot was then asked to provide a subjective rating for each factor by placing an index along a non-ordinal continuum. The TLX scale is a subjective workload analysis tool developed for generic human performance tasks. It breaks workload down into six factors (mental demand, physical demand, temporal demand, performance, effort, and frustration). A numeric score is assigned by the subject for each factor. The factors are rated in order of relative importance to allow weighing of the sub-scale scores to generate a single numeric score (with higher being increased workload). The weighting of the factors is mechanized as a set of 15 "flash cards" where the pilot selected which of the two factors shown are more important. The pilot asked to generate a weighting for each of the tasks flown and assign workload ratings for each test point flown. The factor weight and factor value was then algorithmically combined to report a score given by a numerical value between 0 and 100. Considerations to be made by the subject in each Factor area:

- Mental Demand (high/low) 1. How much mental and perceptual activity was required? (e.g. Thinking, deciding, calculating, remembering, looking, searching, etc.) 2. Was the task easy or demanding, simple or complex, forgiving or exacting?
- Physical Demand (high/low) 1. How much physical activity was required? (e.g. Pushing, pulling, turning, controlling, activating) 2. Was the task easy or demanding, slow or brisk, slack or strenuous, restful or laborious?
- Temporal Demand (high/low) 1. How much time pressure did you feel due to the rate or pace at which the tasks or task elements occurred? 2. Was the pace slow and leisurely or rapid and frantic?
- Performance (Good/Poor) 1. How successful do you think you were in accomplishing the goals of the task by the experimenter (or yourself)? 2. How satisfied were you with your performance in accomplishing these goals?
- Effort (High/Low) 1. How hard did you have to work (mentally and physically) to accomplish your level of performance?
- Frustration level (High/Low) 1. How insecure, discouraged, irritated, stressed, and annoyed vs secure, gratified, content, relaxed, and complacent did you feel during the task?

Workload Rating Scales: Cooper-Harper (CHR)

The traditional Cooper-Harper rating scale was the third and final assessment of pilot workload and served as a common baseline, both to link the new data to the original study (Ref. 6) and relate the POWS and TLX ratings to a familiar reference. The CHR scale is the classical measure by which handling qualities tasks are evaluated. The scale is bi-modal, in that it considers aspects of performance as well as workload. For tasks where desired and adequate tolerances are clearly defined, the pilot was asked to evaluate the test point in accordance with the CHR scale, shown below as Figure 8.

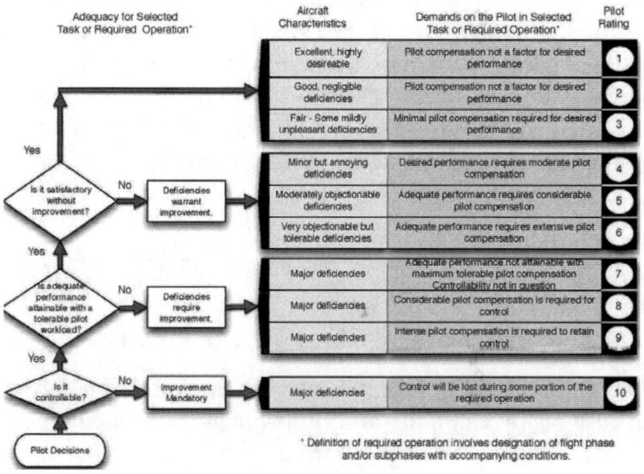

Figure 8: CHR Scale.

Data Collection

Pilots were given an introductory brief on the study's purpose, the general execution strategy for the evaluations, and the rating scales in use. At the start of each task, they were permitted to perform practice runs to gain familiarization prior to conducting formal evaluations. Additionally, the baseline model configuration was always the first presentation. Subsequent evaluations were of the various model permutations for a given task and selected at the test director's discretion. The pilot was never informed as to what particular model version they were evaluating. Following the conclusion of each trial, the recoded inceptor data was saved for post processing and the pilot was asked to provide their ratings using the three prescribed scales. The rating order of POWS, TLX, and HQR was intentionally selected as a best practice to avoid cross contamination of pilot opinion between POWS and HQR, both of which are ordinal, 10-point scales but have unique guidelines for their use.

PIW Metrics

All metrics identified in Ref. 6 were computed for the data. Two of the most promising from the earlier work, aggressiveness and mean power frequency, are highlighted herein when the results are presented. The definitions of the metrics are included for completeness.

The aggressiveness metric (from Ref. 5), denoted J in Equation 1, takes input time histories (from time t_0 to t_f) from the pilot lateral and longitudinal inceptor positions. The terms δ_{ap} and δ_{ep} represent the lateral and longitudinal stick inputs after filtering out all content above 2Hz to only retain conscious pilot intent. The terms vectors $\overline{\delta}_{ap}$ and $\overline{\delta}_{ep}$ are generated by low-pass filtering the raw cyclic displacement vectors at 0.25Hz to generate a representation of long-term control activity such as trim changes. The differencing between the inputs representing pilot intent and the long-term control activity ensures that the metric only inflates when pilot inputs are actively regulating the vehicle state in response to disturbances or actively tracking some target. Versions of the aggressiveness metric as presented below and with collective and pedal inputs added were computed. Due to the minimal amount of activity on the collective and pedal for the sum-of-sines MTEs performed, the inclusion or exclusion of these channels had almost no impact on the evaluated value of the metric for those tasks.

$$J = \frac{100}{t_f - t_0} \sum_{\tau = t_0}^{t_f} \left(\frac{|\delta_{ap}(\tau) - \overline{\delta}_{ap}(\tau)|}{\delta_{ap}^{max} - \delta_{ap}^{min}} + \frac{|\delta_{ep}(\tau) - \overline{\delta}_{ep}(\tau)|}{\delta_{ep}^{max} - \delta_{ep}^{min}} \right) \Delta\tau$$

Equation 1: Aggressiveness Metric.

Work performed by Lampton and Klyde in Ref. 4 suggested the use of wavelet transforms to facilitate the computation of cutoff frequency throughout the time history of a maneuver, and proposed a metric deemed power frequency. The coefficients produced from the wavelet transform allow calculation of the power across both frequency and time. Analogous to the calculation of cutoff frequency, the time varying cutoff frequency (ω_c (t)) is found by determining the frequency at which half the total power has accumulated for each time instance in the time-frequency representation. From Ref. 4, the time varying cutoff frequency is computed as shown in Equation 2. In this expression, the PSD curve at each time instance, $G_{\delta\delta}$ (t), is integrated to identify the total area from $\omega = 0$ to infinity, and the cutoff frequency at the current time instant is determined as the frequency where half the total area is captured.

$$\frac{\psi_1^2(t)}{\psi_{tot}^2(t)} = \frac{\frac{1}{2\pi}\int_0^{\omega_c(t)} G_{\delta\delta}(t)d\omega}{\frac{1}{2\pi}\int_0^{\infty} G_{\delta\delta}(t)d\omega} = 0.5$$

Equation 2: Calculation of time-varying cutoff frequency.

The cutoff frequency is scaled by the maximum power over the frequency range at the time instance under consideration. This is performed for all time instances to develop the power frequency (ω_G (t)), and is shown in Equation 3. Although Equation 3 is written for the time-varying case, it may be applied similarly using the discrete Fourier transform data to determine an average power frequency.

$$\omega_G(t) = \omega_c(t) \max_{\omega} G_{\delta\delta}(t)$$

Equation 3: Calculation of time-varying power frequency.

RESULTS

The results from the study are organized into two parts. First, overall correlation of the CHR, POWS, and TLX scales is examined. Second, observations from the data are discussed, with a focus on how the pilot's perceived the degraders they encountered in the testing.

Overall Correlation

A note on the data presented in the figures examining correlation that follow is warranted. Metrics, identified in Ref. 6 were computed based on the study data. Quality of correlation is desired between the CHR, POWS, and TLX scale to assess the ability of the quantitative metrics to serve as a measure of workload. To that end, some statistical analysis is performed. Most metrics are plotted along with the correlation coefficient computed assuming a best-fit line of the form $\hat{y} = \widehat{B_0} + \widehat{B_1}x$, and an associated prediction band. The coefficients of the best-fit line were selected to minimize the least square errors between the fit value and the data from an individual pilot. The 100(1-α) prediction band for the quantity $\widehat{B_0} + \widehat{B_1}x$ is given as:

$$\widehat{B_0} + \widehat{B_1}x \pm t_{n-2,\alpha/2} \cdot s_{pred} \text{ , where}$$

$$s_{pred} = s\sqrt{1 + \frac{1}{n} + \frac{(x - \bar{x})^2}{\sum_{i=1}^{n}(x_i - \bar{x})^2}}$$

Equation 4: Prediction band calculations.

In Equation 4, $t_{n-2,\alpha/2}$ comes from the t-distribution as a function of the number of degrees of freedom in the dataset and the desired level of confidence in the prediction band, and s represents the standard deviation of the error. The prediction bands included in the following plots are for a 95% confidence level. Thus, for a given value of an evaluated metric, an estimate of the qualitative workload metric is provided by the corresonding value of the best-fit line, and a prediction interval is bounded by the prediction bands to a 95% confidence level.

The linear form of the best-fit line represents an assumption that the PIW metric increases linearly with workload. This assumption is potentially problematic, especially with the CHR scale considering the nonlinear, or perhaps piecewise linear, nature of the scale with regards to workload. The POWS and TLX scales may or may not be interpreted linearly when evaluations are given by the pilot. The use of a more complex fit line may turn out to be appropriate; however, there was a desire not to assume a complicated underlying model structure without first principles knowledge of the relationship present in the data. What was desired, was a simple interpretation to compare the correlation between the PIW metric and qualitative scale, and this is provide directly by the correlation coefficient.

Assuming a linear model form for qualitative handling quality ratings is not unprecedented. Ref. 12, for example, successfully demonstrated nearly monotonic increases in evaluated mean and max power frequency as a function of time with HQR rating for two pilots. Thus, the use of linear statistical methods provides rough measures of correlation to assess the ability of PIW metrics to stand in as an analog to qualitative workload evaluations.

The design of the study was influenced by the results and discussion presented in Ref. 6. The expectation going into the data analysis was that the POWS scale and perhaps the NASA TLX ratings would correlate to PIW measurements better than traditional CHR workload ratings due to how CHR ratings merge disparate assessments (workload and performance) in the process of generating an overall workload assessment. It is noted that NASA TLX also combines various aspects of workload, but for this study the relative weight of each input factor remains the same as the subject matter experts are exposed to degraders.

Overall correlation was first investigated using the most clinical task data available – the single axis sum-of-sines. The actual results have revealed that the correlation is highly dependent on many factors, as the results do not consistently indicate one scale is preferred compared to others.

Consider first single axis sum-of-sines results shown in Figure 9, generated by Pilot 1001. A legend to assist in interpreting the correlation figures is shown as Figure 10. Figure 9 plots the qualitative workload rating versus the PIW metric in various markers (representing specific nominal or degraded configurations), along with a bold best fit line and the dashed line representing the prediction band to a 95% confidence level. Reasonable, but perhaps not as strong as desired correlation is present between the PIW metric and all three qualitative workload ratings. The difference between HQR, POWS, and TLX results is not stark. In fact, approximately the same correlation coefficient and spread in the prediction band is present. Further, the quantitative data (aggressiveness in this case) is spread relatively evenly across a large band promoting confidence in the existence of the correlation.

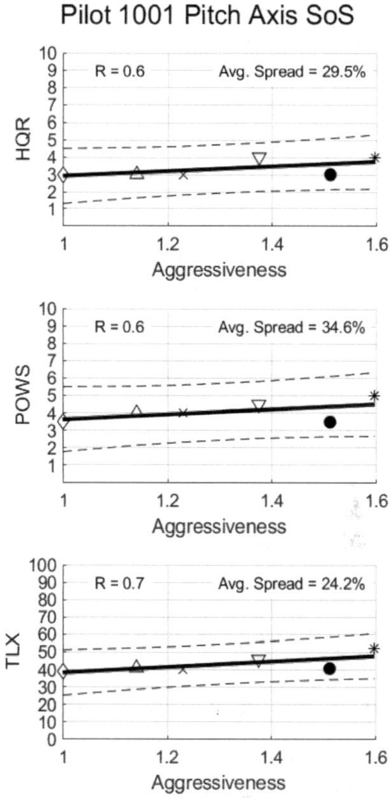

Figure 9: Pitch SoS Correlation with CHR (top), POWS (middle), and TLX (bottom).

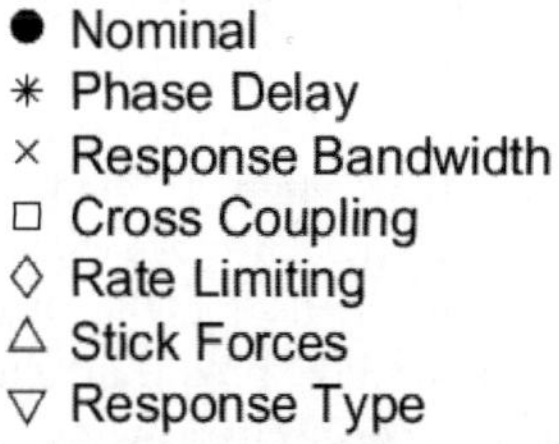

Figure 10: Legend for Interpreting Correlation Figures.

The similar character of correlation among qualitative workload scales is generally not apparent in other pilot data, conducting the same single axis sum-of-sines task. Consider the Pilot 1012 data, representing the most common interpretation of the data we collected in Figure 11:

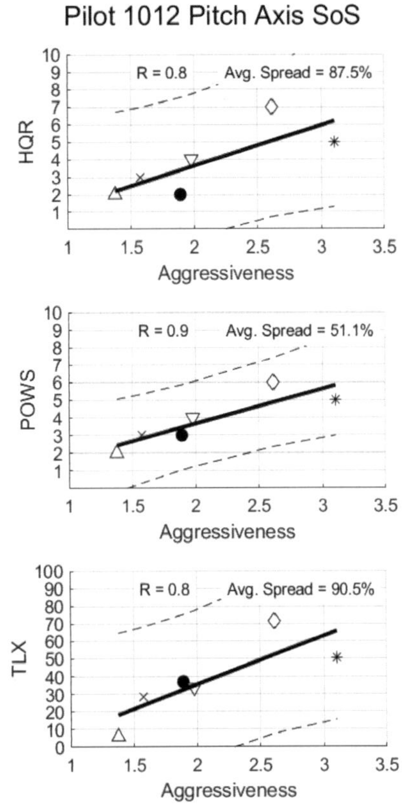

Figure 11: Pilot 1012, Pitch Sum-of-Sines Data.

The Pilot 1012 data shows the trends apparent in 8 out of 12 cases for pitch sum-of-sines task data – the POWS correlation with PIW metrics provides the lowest spread, therefore best predictive capability, followed by the CHR comparison. The predictive capability for the TLX data is worse than the other qualitative metrics for 11 out of the 12 pilot's data. It is not uncommon to see TLX predictive bands spreading over 75% of the TLX scale range, effectively nulling any predictive capability based on PIW metrics as the input. The poor predictive capability of TLX when associated with a handling qualities task combined with the fact that this scale is the most difficult to implement dim the prospect of using this scale in future experiments related specifically to handling qualities.

Overall, correlation measured by correlation coefficient was poorer than anticipated for the PIW metrics of aggressiveness and mean power frequency, with wide ranges seen across the pilot population. Table 4 summarizes the correlation coefficients computed by comparing the qualitative workload ratings against the aggressiveness metric for the pitch only SoS task. In general, most subjects aligned exhibited a positive correlation, albeit to varying degrees, which aligned with previous expectations. With closer inspection, several negative values appear which was unanticipated. One may be quick to dismiss the results of pilot 1010 since their flight experience was quite limited, however pilot 1011 exhibited similar traits and was an active Marine test pilot with over 1800 flight hours. This incongruence challenges the premise that all pilots materialize workload in a qualitatively similar fashion and demonstrates the challenge in seeking the universal application of a PIW metric.

Table 4: Correlation Coefficient Summary – Pitch SoS Task for Aggressiveness PIW Metric.

Evaluator	R - CHR	R - POWS	R- TLX
1001	0.6	0.6	0.7
1002	0.6	0.4	0.4
1003	0.6	0	0.1
1004	0.8	0.9	0.3
1005	0.8	0.8	0.7
1006	-0.3	0	0.4
1007	0.3	0.9	0.7
1008	N/A	0.5	0.1
1009	0.7	0.7	0.6
1010	-0.6	-0.6	-0.6
1011	N/A	-0.3	-0.5
1012	0.8	0.9	0.8

Following the results and conclusions from the first survey of PIW metrics (Ref. 6), it was identified that correlation would likely need to be specific per-pilot, or at least to "types" of pilots. The hypothesis was that with data collected for purpose coupled with fitting each, individual pilot, a sufficiently strong correlation could be established to allow application of PIW metrics as a more regular tool during HQ evaluations in lieu of subjective ratings. The data suggests that, while it seems that considering individual pilot correlations is necessary, it is not sufficient with PIW metrics. The results show that some individual pilots demonstrated strong correlation, while some pilots demonstrated no identifiable correlation at all. This does not necessarily indicate that a given pilot is objective in their assigned workload ratings. It only means that the workload perceived by some pilots don't significantly overlap with the quantitative measures of workload attempted to be captured by the PIW metrics. It suggests that different metrics should be investigated as a standalone, or combinatorial measure of workload perhaps relying on additional data, such as physiological measures, which were not captured in this study.

Effect of the Degraders – Sum-of-Sines Tasks

In a perfectly designed experiment, the baseline CHR configuration would have always received a Level 1 qualitative rating, meaning the pilot was able to achieve desired task performance by modulating perceived workload. Pilot 1012's ratings are selected to highlight how close to this ideal the experiment came.

The baseline model characteristics were designed using the quantitative criteria in ADS-33E to provide borderline level 1/level 2 HQ. For the single-axis SoS MTEs, this was found largely to be the case as shown in Figure 12. When examining this figure, and similar examples that follow, the qualitative ratings for each degrader are shown as black bars, with the nominal configuration indicated by the red line. The intent is that this allows a quick look at relative difficulty, as perceived by the pilot. That said, there was a marked increase in both HQR and POWS associated with the multi-axis SoS MTE as show in Figure 13. Not only does the increased task complexity increase the target baseline, but it also seems to exaggerate the effect of the degraders. Potentially because the pilot is starting to run out of capacity to deal with degraded modes and the harder task (i.e. the pilot is becoming task saturated).

For the majority of the pilots, the degraders tended to result in worse HQRs and higher workloads than the baseline configuration. For a couple of the pilots, this was not the case as exemplified in Figure 14. This could be due to a learning effect as the pilots were shown the baseline configuration first (though they did not know it was baseline). This learning effect occurred even though pilots were encouraged to take as many practices runs as necessary to feel competent in the task.

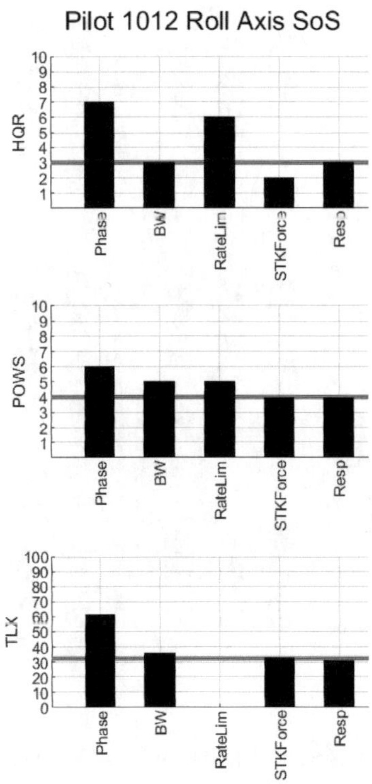

Figure 12: Pilot 1012 Roll Sum-of-Sines Pilot Ratings.

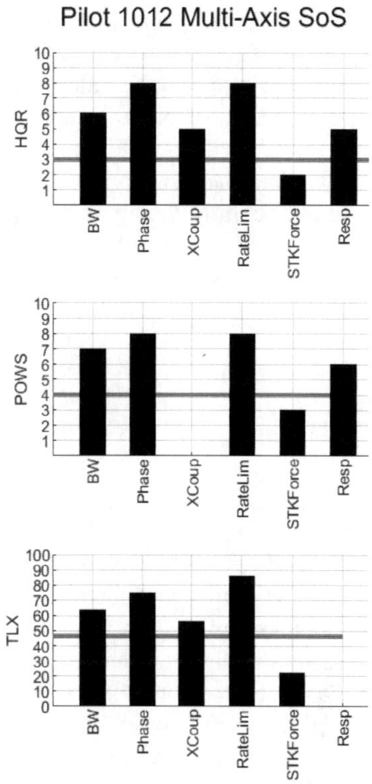

Figure 13: Pilot 1012 Multi-Axis Sum-of-Sines Pilot Ratings.

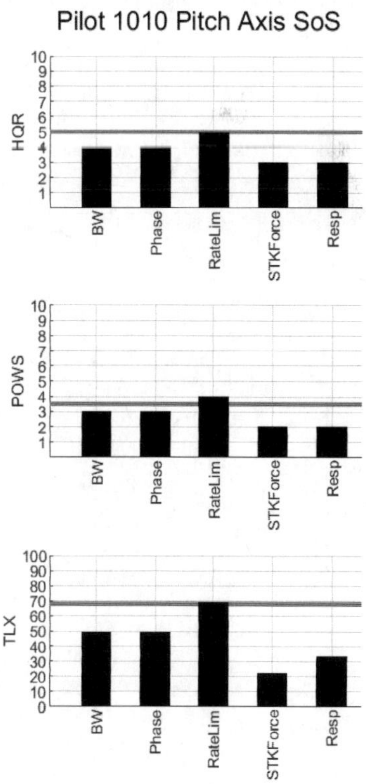

Figure 14: Pilot 1010 Qualitative Ratings for Pitch Axis SoS.

For the case of the attitude command system, the effect of the model change was intended to improve the handling qualities by augmenting the traditional rate command system. Interestingly, a majority of pilots found this configuration more objectionable, as shown in Figure 13. The pilots often made comments that the sensitivity was too high which caused them to over control the aircraft. One potential cause for these responses may include the pilot's adaptation to a different response type, in which case the study accommodated for this by permitting pilot's to conduct proficiency run prior to scoring, although subjects rarely capitalized on this opportunity. A second cause may have been the gearing ratio between the inceptor and attitude output.

An expected result is noted when considering the relative location of the nominal configuration and phase delayed configuration in terms of the PIW metrics and pilot ratings. In a closed-loop compensatory tracking task, the phase delay is expected to result in higher PIW measurements due to the need to over apply the input to illicit an initial response and then continue to over control to drive the response to the desired set point. Most often when considered on a per-pilot basis, the qualitative and quantitative data did show this expected trend. A sample case is seen in Figure 15, where the phase delayed configuration is in the upper right-hand corner of the plot for all qualitative scales, maximizing the PIW metric of mean time-varying power frequency in the longitudinal control axis.

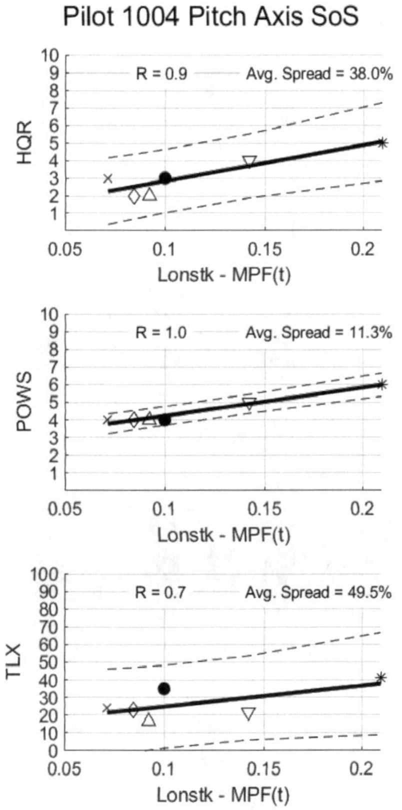

Figure 15: Pilot 1004 Exhibiting Max Qualitative/Quantitative Workload in Phase Delayed Configuration.

Effect of the Degraders – Precision Hover Task

Elaborating on the expectation that the baseline model was Level 1, it stands to reason that the first five degraders (Stability, Phase delay, Bandwidth, cross couplings, and rate limit) would yield ratings that are higher than baseline. Furthermore, the modified response type was expected to reduce pilot workload thus leading to ratings equal to or lower than baseline. Increasing the stick force was initially expected to degrade handling qualities, but in the course of collecting data, it was generally considered by pilots as an improvement as it provided a sense of damping when they might otherwise tend to over control the aircraft. The precision hover was the most challenging of the five tasks, so much so that two pilots were incapable of completing it for record. Of the remaining pilots that were successful, two pilots rated the baseline at Level 1, six pilots as Level 2, and finally two pilots as Level 3. Figure 16 shows the precision hover results for Pilot 1001 which best exemplifies the expected shape of the data.

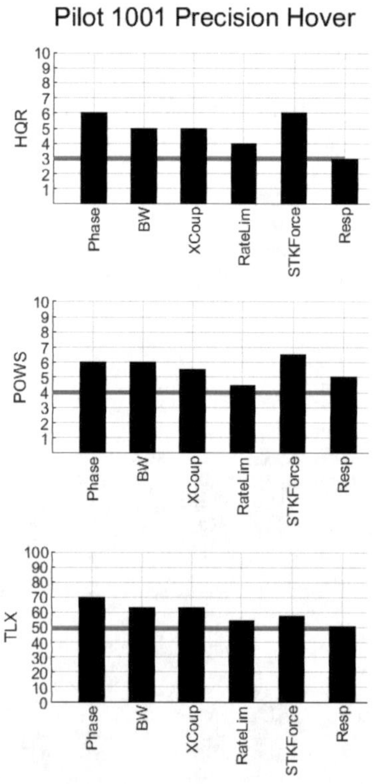

Figure 16: Pilot 1 Data shape Exemplified.

The modified response type was expected to reduce pilot workload, and it was generally an improvement resulting in an HQR reduction of approximately 2 from the baseline assessment. This was in contrast to the SoS result in which the modified response type did not markedly reduce workload. This inconsistency can be attributed to the fundamental difference in the style of tracking employed for the Precision Hover MTE as compared to the SoS task.

A well know challenge in the pursuit of assessing handing qualities is achieving a true assessment at the hands of a focused, proficient, and competent pilot. The learning curve associated with undertaking an unfamiliar task must be managed so as prevent unfairly inflated pilot ratings. Despite best efforts, some pilots insisted on their readiness to collect data before reaching full proficiency. This is demonstrated in the results for pilot 11 which are shown in Figure 17. Their baseline assessment was exceptionally high on all workload scales. Subsequent events in which the degraders were applied indicate a slight reduction in the HQRs but overall still high workload according to the alternate scales. Performance peaked during the assessment of stick force and response type, initially breaking into the level 2 category before ultimately giving an HQR 4 rating and achieving level 1 standards. This marked improvement was the synthesis of repetition and favorable model settings.

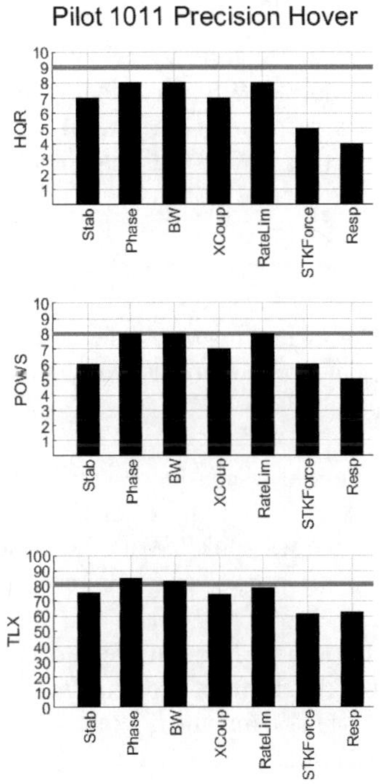

Figure 17: Pilot 11 gaining proficiency through repetition.

Overall, the workload ratings across all scales investigated were generally consistent for the population tested. Pilot 7 however presented with somewhat conflicting ratings, in particular between HQR and POWS, as seen in Figure 18. Their HQRs averaged approximately 4 while their POWS ratings were near 7. This pilot found themselves in a quandary as they were routinely working extremely hard and achieving desired performance. In practice, it's far more typical to observe pilots complacent to achieve adequate tolerance at a much lower workload and thus motivate them to seek desired performance. In this pilot's situation, it was the structure of the HQR scale that drove them to select a rating consistent with their performance level even though the rating failed to capture the workload necessary to reach that threshold. This fundamental constraint of the HQR scale was the motivation to explore the suitability of alternate workload scales that may better correlate PIW metrics to perceptions of workload.

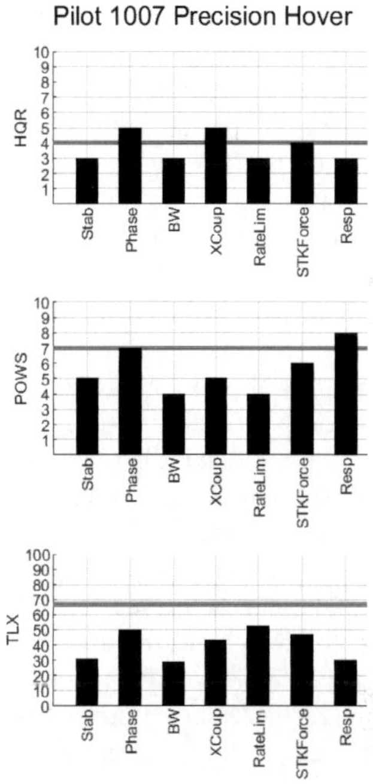

Figure 18: Pilot 7 Rating variation across scales.

Effect of the Degraders – Reposition Task

For the reposition task, only POWS and TLX qualitative rating data was collected due to the lack of defined tolerances for the open-to-interpretation task. The reposition task data did not add value to the attempt to find correlation between the PIW metric and qualitative handling quality rating.

Of note, we saw little variation in the nominal configuration workload for the evaluation pilots, where POWS range remained quite limited from 2-4. The effect of the degraders was more pronounced for some pilots than others. Some pilots kept POWS ratings from 2-6 for all degraders, whereas in the extreme case we saw Pilot 1009 rating the bandwidth degrader a POWS 10, although this did not comport with the corresponding TLX rating of 63, which was exceeded by other degraders on the TLX scale.

Experimental Data Collection Lessons Learned

In executing the experiment, several observations were noted that would greatly benefit future researchers interested in recreating this type of experiment. The criticality of the simulator field of view is a function of the task to be completed. Two of the tasks in this experiment required the pilot to fly typical dynamic helicopter maneuvers. While the reposition task had only a few key requirements to meet, the hover MTE was highly specified in time and distances and required extreme precision to perform to the desired tolerances. The limited field of view of the single seat simulator at USNTPS, which was an approximately 8 feet by 8 feet flat projection on a wall approximately 10 feet in front of the pilot, did not permit the subject to view any of boresight cues critical to accomplishing the Hover MTE which severely restricted their ability to accurately perform the task and assess quality. Guidance from the test conductor on how to use the available cueing off the nose to assist in managing fore/aft position was used, however this cue was highly sensitive to the air vehicle pitch attitude.

Recurring pilot comments such as "I can't make an input small enough" and "I feel a heavy clunk as I move the cyclic stick through trim" highlighted an influence of the flight control system on the pilot's ability to perform some tasks. An audible squeal in the roll axis also indicated that potential bearing wear and friction in the control mechanism was contributing to the overall experience. It was apparent that the flight control system mechanical characteristics were having an influence on the pilot workload with the effect most apparent during high precision tracking when stick amplitudes were very small as in the positon keeping portion of the precision hover MTE or certain portions of the SoS task. Given that all subjects were exposed to the same flight control system, the data are expected to be comparable across the sample group, although this variable was not an intended source of degradation to inflate pilot workload. Correlation is expected to still be viable, although the magnitude relationship of the metric to the qualitative rating may be biased higher than if the flight control system was optimally designed.

Pilot Proficiency

Observations from the simulation sessions overwhelmingly substantiate the notion of a proficiency curve with respect to conducting a task. While the pilot was afforded the opportunity to perform each new task to become familiar, it was typical for them to accept only one or two practice runs prior to taking data for score. This element did not factor heavily in the simplified SoS tracking, but became strongly apparent with the much more challenging reposition and hover tasks. The learning curve can be discretized into three sections. In the first section, the pilot is translating their understanding of the task into control inputs to perform the task. Task performance typically improves demonstrably after 3-5 attempts at which point it plateaus. This period continues until fatigue and/or frustration causes performance to degrade with further attempts. Each pilot's curve is unique, but the overall trend appears consistent. The pilot's reluctance to conduct additional proficiency runs prior to taking workload data for the complicated tasks meant the pilot was still climbing the initial learning curve before reaching their individual plateau and resulted in artificially inflated CHR. While this alone may not be significant, the order in which the degraders were presented to the pilot will need to be carefully examined to prevent false conclusions that the degraders presented early in the sequence were in fact harder to fly when the reality may be that pilots were still gaining proficiency through those runs. This was verified in one pilot's case who, when complete with all degraders for the hover MTE, was asked to repeat a previously objectionable case in which vehicle control was in question and performed the task safely. The scope of data collection averaged 3 to 4.5 hours per subject and likely contributed to some pilot's unwillingness to accommodate their natural learning curve for each task.

CONCLUSIONS

For the purposes of correlation with PIW metrics, it seems that neither CHR nor POWS was notably better than one another. This was surprising since CHR is not solely a workload scale whereas POWS was intended to be.

For the purposes of correlation with PIW metrics, it seems that TLX is consistently less correlatory than either CHR or POWS. This does not necessarily mean TLX isn't an appropriate tool for evaluating workload in an HQ sense, just that it doesn't emphasize the form of workload being captured by PIW metrics.

It is becoming more evident that establishing quantitative correlation of workload will require consideration of pilots as individuals but that, at least for PIW metrics, not all pilots can be expected to show strong, or even any, correlation.

In cases where the most offending degraders did not illicit objectionable workload, pilot control strategies did not sufficiently stress the system which reinforces the importance of using multiple pilots to assess the suitability of a flight control system.

Future experimental design should consider the end to end performance of the system and the desired effect verified by a test pilot.

Future research should consider reducing scope, breaking sessions into multiple events, and forcing additional practice runs to fully develop pilot proficiency prior to collecting data.

Author contact: Ryan Paul ryan.paul@okstate.edu
Matthew Rhinehart matthew.h.rhinehart2.civ@us.navy.mil

Kenneth Cahill kenneth.m.cahill2.civ@us.navy.mil
Zachary Smith Zachary.f.smith6.civ@us.navy.mil

ACKNOWLEDGMENTS

R. C. Paul thanks the ONR Summer Faculty Program for the support provided to perform the analysis portion of this work. The team would like to thank the NAWCAD NISE Office for being the primary funders and backers of this research. In addition, a thank you to all those involved in the simulation test; especially the pilots who performed the flight tests and US Navy Test Pilot School for the use of their facilities.

REFERENCES

[1] S. Hart and L. Staveland, "Development of NASA-TLX (Task Load Index): Results of Empirical and Theoretical Research,," *Advances in Psychology,,* vol. 52, pp. 139-183, 1988.

[2] R. Paul and M. Rhinehart, "Exploring Pilot Workload Using Inceptor Time Histories," in *Vertical Flight Society 76th Annual Forum and Technology Display*, Virtual, 2020.

[3] A. Lampton and D. Klyde, "Power Frequency: A Metric for Analyzing Pilot-in-the-Loop Flying Tasks," *Journal of Guidance, Control, and Dynamics,* vol. 35, no. September-October 2012, pp. 1526-1537, 2012.

[4] G. E. Cooper and J. R. P. Harper, "The Use of Pilot Ratings in the Evaluation of Aircraft Handling Qualities," National Aeronautics and Space Administration, Washington, D.C., 1969.

[5] D. Wilson and D. Riley, "Cooper-Harper Pilot Rating Variability," AIAA Atmospheric Flight Mechanics Conference, Boston, MA, 1989.

[6] D. Riley and D. Wilson, "More on Cooper-Harper pilot rating variability," in *17th Atmospheric Flight Mechanics Conference* , Portland, OR, USA, 1990.

[7] United States Army Aviation and Missle Command, *ADS-33E-PRF Aeronautical Design Standard Performance Specification Handling Qualities Requirements for Military Rotorcraft,* Redstone Arsenal, AL: Army Aviation Engineering Directorate, 2000.

[8] C. Hanson, J. Schaefer, J. J. Burken, D. Larson and M. Johnson, "Complexity and Pilot Workload Metrics for the Evaluation of Adaptive Flight Controls on a Full Scale Piloted Aircraft," National Aeronautics and Space Administration, Dryden Flight Research Center, Edwards, CA, 2014.

[9] E. J. Field and S. E. Giese, "Appraisal of Several Pilot Control Activity Measures," AIAA Atmospheric Flight Mechanics Conference, San Francisco, CA, 2005.

[10] S. Jennings, G. Craig, S. Carignan, K. Ellis and D. Thorndycraft, "Evaluating Control Activity as a Measure of Workload in Flight Test," *Proceedings of the Human Factors and Ergonomics Society Annual Meeting,* vol. 49, no. 1, p. 64–67, 2005.

[11] M. B. Tischler, Aircraft and Rotorcraft System Identification, Reston, VA: American Institute of Aeronautics and Astronautics, Inc., 2012.

[12] S. A. Conner and W. W. Wierwille, "Comparative Evaluation of Twenty Pilot Workload Assessment Measures Using a Psychomotor Task In A Moving Base Aircraft Simulator," National Aeronautics and Space Administration, Moffett Field, CA, 1983.

[13] I. Niewind, "Pilot Gain and the Workload Buildup Flight Test Technique: A Closer Investigation of Pilot Inceptor Workload," German Aerospace Center, Braunschweig/Manching, 2012.

[14] I. Niewind, "Pilot Gain and the Workload Buildup Flight Test Technique: A Closer

Investigation of Pilot Inceptor Workload," German Aerospace Center, 2012.

[15] D. Klyde, S. Pitoniak, P. C. Schulze, P. Ruckel, J. Rigsby, C. Fegely, H. Xin, W. Fell, R. Brewer, F. Conway, R. Mulato, J. Horn, C. Ott and C. Blanken, "Piloted Simulation Evaluation of Tracking Mission Task Elements for the Assessment of High-Speed Handling Qualities," *Journal of the American Helicopter Society,* vol. 65, no. 032010, pp. 1-24, 2020.

[16] D. T. McRuer, D. Graham and a. I. Ashkenas, Aircraft dynamics and automatic control, Princeton University Press, 2014.

[17] W. Navidi, Statistics for Engineers and Scientists, New York, NY: McGraw Hill, 2008.

On the Characterization of Nonlinear Frequency Interactions of the Pilot Vehicle System (Rotorcraft): A Preliminary Study[1]

Robert Alstrom PhD
Research Engineer
Vertical Lift Research Center of Excellence
Georgia Institute of Technology
Atlanta GA, USA

Ryan Paul PhD
Assistant Professor
Mechanical and Aerospace Engineering Department
Oklahoma State University
Stillwater, OK, USA

INTRODUCTION

To begin this study on the identification of nonlinear frequency interactions (or quadratic phase coupling) in the pilot-vehicle system, we present two examples from previous research are highlighted. The first example comes from a flight experiment that was conducted using the NRC Bell 205 Airborne Simulator to study the effect of an under-damped engine governor on the vertical axis handling qualities. The governor damping ratios and natural frequencies were varied and flown/evaluated using several pilots. For the engine rotor configurations with low damping ratios, the evaluation pilots noted an increase in the engine torque oscillations as well requiring the employment of different collective stick compensation strategies to suppress the resultant oscillations. For the resonant prone engine-rotor configurations, the torque dynamics were easily excited by nominal collective stick displacements; this experimental result would suggest that the frequency band (~0.6 rad/s to 2.5 rad/s) of the pilots' collective stick activity is coupling nonlinearly with the resonant dynamics of the prescribed engine-rotor configurations (Ref.1). The second motivating example comes from Smith and Edwards (Ref. 2), which documents the design of a pilot induced oscillation suppression (PIOs) filter. The PIO filter contains an adaptive element of the pilots stick gearing schedule which operated by sensing the frequency and amplitude of the pilots stick displacements. The PIOs filter will change the *shape* of the pilots input stick displacements; by changing the shape of the stick input, one is effectively modifying the frequency content of the stick input in a nonstationary manner. This implies that there is a nonlinear frequency coupling between the vehicle dynamics and the pilot input; thus, by shaping the pilot's input, it is possible to inhibit the interaction of certain input Fourier components with the ringing frequency (~3.5 rad/s) in the longitudinal dynamics of the vehicle. Pavel and her colleagues (Ref. 3) write extensively on the topic adverse rotorcraft pilot couplings (A/RPCs). They give a comprehensive review of the differences and similarities between fixed and rotary wing couplings, they provide definitions, historical aircraft incidents of rotorcraft pilot coupling and the state of the art on modeling, simulation and analysis of rotorcraft pilot couplings, however there is no discussion on the effect of nonlinear frequency interaction on rotorcraft pilot coupling.

Identifying frequency coupling effects and the degree of nonlinear coupling is achieved with the application of Higher Order Spectral Analysis (HOSA) on either experimental or modeled data. Perhaps where HOSA has been most successfully applied is in the field of aeroelasticity and structural dynamics. Aeroelastic phenomena include unsteady nonlinear aerodynamic loadings, structural nonlinearities and the coupling between the flow and the structural response. Chabalko (Ref. 4) and Silva (Ref. 5) have applied this advanced signal processing technique to experimental data from the High-Speed Civil Transport (HSCT) wind tunnel experiment and the F-16 wing/stores flutter clearance flight test data. For the HSCT experiment, the results demonstrated that *hard flutter* is related to the intermittent nonlinear coupling between the shock wave displacement and the large amplitude structural response of the flexible semi-span model (FSM). For the F-16 store clearance data, HOSA was used to identify the nonlinear interactions associated with maneuver induced limit cycle oscillations (LCOs). The results of this analysis showed that there was quadratic as well as cubic coupling present in the acceleration signals of the under-wing launchers and high quadratic coupling levels between the flaperon displacements and the under-wing oscillations. In both of these studies, higher order spectra analysis is used primarily as a diagnostic tool to examine *documented flutter* instability and LCO

---

[1] Presented at the VFS Handling Qualities Technical Meeting, Huntsville, Alabama, USA Feb. 7-8, 2023.

incidents. In fact, the further purpose of these two aeroelasticity examples was to demonstrate how higher order moment and cumulant spectra can be used to expose the nonlinear physics that is otherwise hidden and not elucidated because classical signal processing methods do not directly indicate the presence of nonlinear phenomena.

Higher order spectral analysis has also been successfully applied as diagnostic tool in the fields of experimental aerodynamics (Ref .6) and the active control of a nonlinear dynamical system (Ref. 7) and in neuroscience (Ref. 8). To the authors knowledge, higher order spectral analysis has not been directly applied to problems concerning flight dynamics, hence this study is the first step in understanding how the mathematical properties of this advanced signal processing technique can be applied to flight dynamics and flying qualities research. The objectives of this exploratory study are the following: (1) *detect* and *identify* the nonlinear frequency interactions in the prescribed pilot-vehicle system, and (2) to determine what role the identified nonlinear frequency interactions play in the observed dynamics of the pilot-vehicle system. To that end, the HOSA technique is applied to flight simulation data gathered at the US Naval Test Pilot School single seat device. The single seat device features a nominal ADS-33 Level 1 longitudinal handling qualities and those from a test case whereby a time delay non-linearity was inserted to degrade the handling qualities to a level 2/3 boundary case as defined by the ADS-33 standard.

SYSTEM DYNAMICS AND DATA ANALYSIS

System Dynamics

The model used in this analysis is a linear time invariant (LTI) model of a generic helicopter in hover with and without an input delay ($\tau = 0.1s$). The model was implemented in the USNTPS fixed-base flight simulator in the MATLAB/SIMULINK environment. The task is 90 seconds in duration and the sampling frequency is 1 kHz. The pilot's input is generated as a result of the sum of sines task, as specified by Ref. 9. The sum of sines task is typically used to evaluate handling qualities in a tight closed loop tracking task, feel systems and control sensitivities and identify pitch bobble or Pilot Induced Oscillation (PIO) tendencies. The prescribed frequencies are provided in Table 1. This flying task is driven by an automated command signal generated by the simulation environment. The task starts from a steady, wings level disposition. The plant dynamics feature typical characteristics of a helicopter in hover. A quartic linear system describes the longitudinal dynamics, as shown in Equation 1 where the state vector is represented as $x'_{long} \triangleq [u\ w\ q\ \theta]$. An eigenvalue calculation reveals that the aircraft has two real roots in the left half plane, one at 0.0541 Hz (0.34 rad/s), the other at 0.3628 Hz (2.28 rad/s) and an unstable complex conjugate pair $0.0249 \pm j0.412$, where the imaginary component coverts to ~0.066 Hz.

Table 1. Precision Aggressive MTE, sum of sines (Ref. 9)

ω_i^{sos} (rad/s)	f_i^{sos} (Hz)
0.314	0.0499
0.523	0.0830
0.837	0.1332
1.361	0.2166
2.199	0.3499
3.560	0.5660
5.760	0.910

$$A_{long} = \begin{bmatrix} -0.0257 & 0 & 0 & -32.2 \\ 0 & -0.3404 & 0 & 0 \\ 0.0120 & 0 & -2.2 & 0 \\ 0 & 0 & 1 & 0 \end{bmatrix} \qquad [1]$$

Figure 1 shows the time histories of the pilot input signal and the corresponding frequency spectra with/without time delay. Figure 1 (a) shows that the mean of the pilot's input signal varies with time, while Figure1 (c) there are significant no deviations in the mean from equilibrium position. It might be instructive to decompose the pilots input signal for the zero-time delay case to understand what the nature of the slowly varying mean is. In Figure 1 (b), the frequency spectra reveals that there is significant spectral activity between say ~0.05 Hz and ~ 2 Hz with the largest spectral magnitude at ~ 0.9176 Hz (close to the input frequency of 0.91Hz). The reader will observe that there is a bimodal spectral hump ranging from 2 Hz to 6 Hz whose depression is centered at ~ 4 Hz (indicated by the red arrow). In Figure 1 (d) shows the spectra for the time delay of 0.1s. Again, there is spectral activity in the band 0.05 Hz to 2

Hz, with the largest spectral peak at 0.5647 Hz. It is observed that the presence of the time delay on the pilots input causes a shift of the peak frequency to a lower value, in fact the difference in frequency is 0.3529 Hz. We can also observe a spectral hump that is centered on ~4.41 Hz (indicated by the blue arrow). The spectral depression at ~4 Hz is filled in/amplified. The time delay causes amplification at frequencies between 2 Hz and 4 Hz. There is no discernable shift in frequency between 2 and 4 Hz and there is no frequency shift.

Next, we will present the pilot input signal with the simulated aircraft response signal and the corresponding spectra. The frequencies for the input and response spectra for both cases are frequency tagged for ease of identification when interrogating the bicoherence matrices. For completeness, the pilot's input is shown here again with the simulated aircraft response signal and the corresponding spectra, with and without time delay. For the zero-time delay case, Figure 2 (c) shows the aircraft response. The reader will note that there are four large excursions in positive pitch (indicated by the numbered dots) at 12.34s, 25.56s ,49.06s and 71.48s. In the corresponding spectra (Figure 2 (d)), the response bandwidth is ~ 1Hz. Figure 3 (c) shows the simulated aircraft response signal for the system with time delay. Again, it can be observed that there are four large excursions in the negative pitch. They are located at 15.33s, 37.99s, 52.20s and 75.42s. The pitch displacement between 52.20s and 75.42s is observed to be between +0.05 radians and -0.02 radians. The aircraft response bandwidth is again ~1Hz.

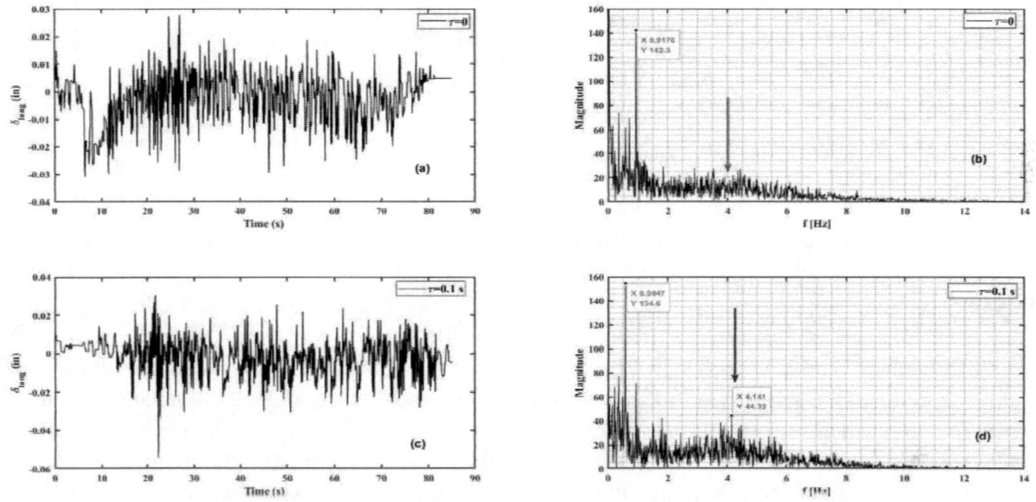

Figure 1. Pilot input signal and frequency spectra (full range)

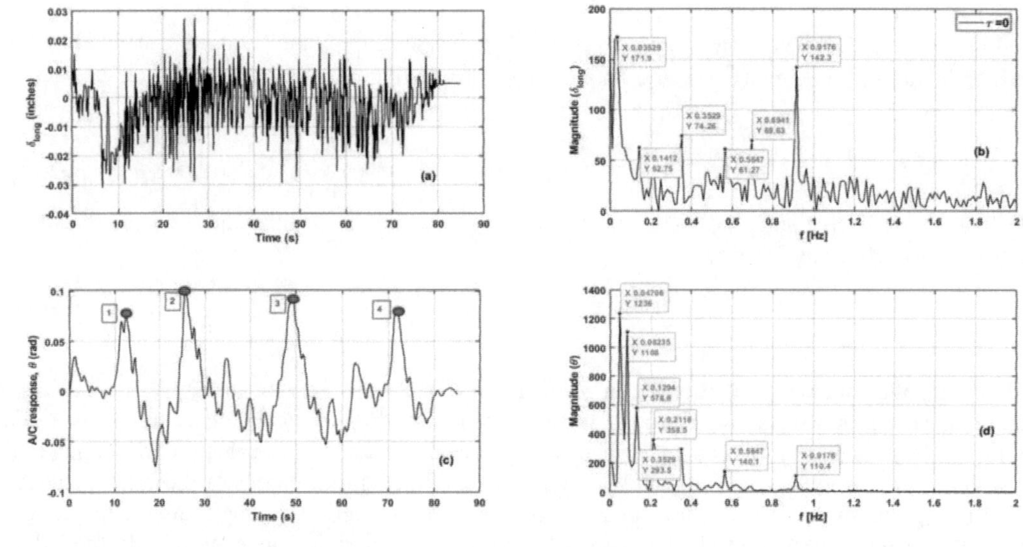

Figure 2. Pilot input and aircraft response signals with frequency spectra ($\tau = 0$)

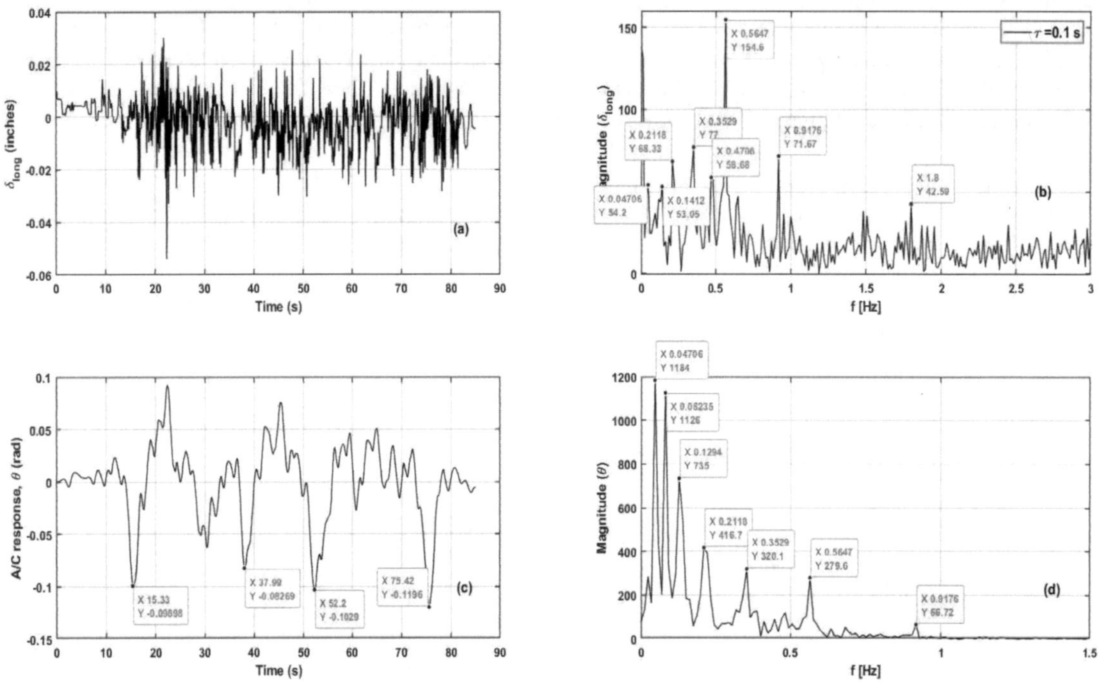

Figure 3. Pilot input and aircraft response signals with frequency spectra ($\tau = 0.1s$)

Data Analysis Method

In general, bicoherence is an advanced signal processing technique that can identify nonlinear phase coupling (or quadratic phase coupling-QPC) between frequency components and the strength of that coupling in either an individual signal (autobicoherence) or multiple signals, namely two (cross-bicoherence). In this project, we will use the *cross-bicoherence* calculation in order to identify the quadratic phase couplings between the multi-harmonic pilot input and the simulated aircraft response signals. The continuous wavelet transform has logarithmic frequency resolution. The wavelet transform of a function s(t) is defined as:

$$W(\xi,\tau) = \int_{-\infty}^{+\infty} s(t)\psi^*_{\xi,\tau}(t)dt \qquad [2]$$

Where ξ and τ are scale and time variables respectively; $\psi^*_{\xi,\tau}(t)$ represents the wavelet family generated by continuous translations and dilations of the mother wavelet. The translations and dilations are obtained by:

$$\psi_{\xi,\tau} = \frac{1}{\sqrt{\xi}}\psi\left(\frac{t-\tau}{\xi}\right) \qquad [3]$$

The wavelet to be used for this analysis is the log-normal wavelet which has logarithmic frequency resolution. The log-normal wavelet is defined as follows:

$$\psi(t) = e^{-\frac{2\pi f_o \log t}{2}} \qquad [4]$$

In this definition, the selection of f_o is based on whether one wishes to capture low or high frequency components in the signal or signals. This wavelet transform is the basis for the time-frequency and bispectral analysis. Quadratic Phase Coupling (QPC) is said to have occurred when two frequencies f_1 and f_2, are present in the signal at the same time along with the sum (or difference) of the two frequencies and the sum of phases of these frequencies remains constant. The magnitude of the bicoherence ranges between zero and unity; complete quadratic phase-coupling happens when the bicoherence is one or closed to one. The two harmonic components are decoupled when the bicoherence magnitude is close to zero or equal to zero. If the nonlinearity is greater than order 2, then it is also possible to obtain zero magnitude for the bicoherence. The cross-bicoherence and is defined as follows:

$$\left|b_{yxx}^{w}(\xi_1,\xi_2)\right|^2 = \frac{\left|B_{yxx}^{W}(\xi_1,\xi_2)\right|^2}{\left(\int_T \left|W_x(\xi_1,\tau)W_y(\xi_2,\tau)\right|^2 d\tau\right)\left(\int_T \left|W_y(\xi,\tau)\right|^2 d\tau\right)} \quad [5]$$

The bicoherence calculation produces a square matrix that is usually presented as a contour map. The contours change shape as a result of some parameter variation. The contours spread or elongate to incorporate motions at other frequencies. The contour map is organized around the main diagonal of the matrix, and the off-diagonal regions of the matrix. The main diagonal indicates *quadratic phase self-coupling (QPSC)*, $\left[b_{yxx}^{w}(\xi_1 = \xi_2)\right]^2$; it is a measure of the interaction between components of the same frequency. The contours in the off-diagonal regions of the matrix indicate the level of cross-frequency coupling in a signal or two signals (cross-bicoherence calculation). There are nine indices of bicoherence, the index of bicoherence that may be useful for this preliminary analysis is the summed bicoherence. The summed bicoherence as a function of frequency indicates the distribution of quadratic phase coupling strength across frequencies. The summed bicoherence is given as:

$$\sum b^2(\xi) = \frac{1}{N_\xi}\sum b^2(\xi_1,\xi_2) \quad [6]$$

N_ξ is the number of frequency components. The phrases nonlinear modal interaction, quadratic phase-coupling and nonlinear frequency interaction will be used interchangeably throughout this work.

A SAMPLE CALCULATION: RESULTS AND DISCUSSION

Case 1: $\tau = 0$

Figure 1 (a), (b) and (c) show the cross-bicoherence matrix, the diagonal (indicated by the white dashed line) and summed bicoherence spectra respectively. To make the matrix easy to read and inspect, an annotated cross-bicoherence matrix is provided in Figure 5.

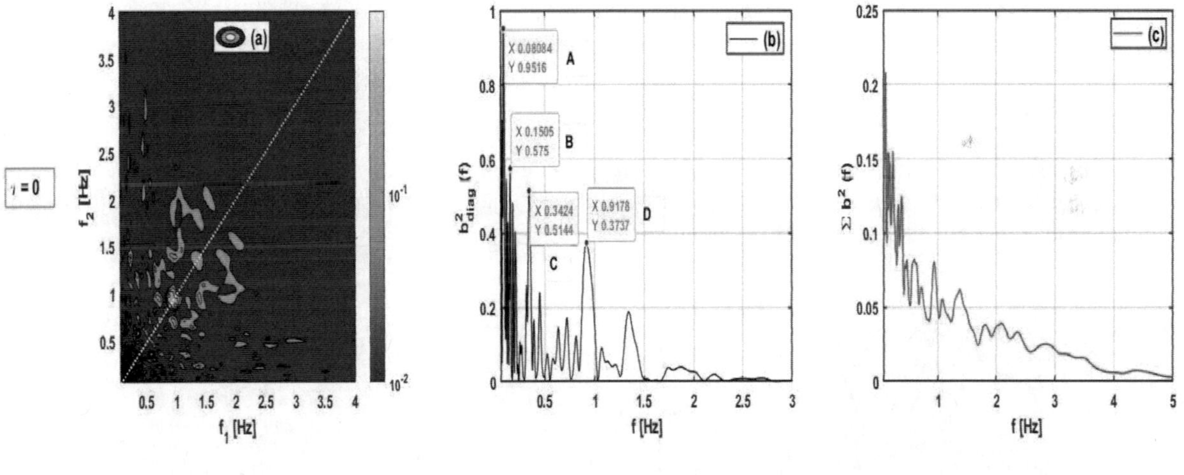

Figure 4. Cross-bicoherence matrix and spectra ($\tau = 0$)

The matrix is organized around the main diagonal and the off-diagonal regions of the matrix. Quadratic phase self-coupling is a measure of the interaction between frequency components of the same frequency. In Figure 5, there are three matrix spectral features indicated, two of which are off diagonal. At Contour A $[b^2(0.2183\text{ Hz}, 0.3498\text{ Hz}) = 0.729]$; this frequency pair has significant nonlinear coupling. The first entry in the frequency pair, 0.2183 Hz is closest to the input frequency 0.2166 Hz and the second, 0.3498 Hz is essentially matched to the input frequency 0.3499 Hz. The resultant frequency, $f_3 = f_1 + f_2$ is 0.5681 Hz which differs from the input frequency 0.5660 Hz by 0.0021Hz. At Contour B $[b^2(0.3523\text{ Hz}, 0.5727\text{ Hz}) = 0.54]$, the frequency pair is also strongly coupled and are close to the input frequencies 0.3499 Hz and 0.5660 Hz respectively. Along the main diagonal at low frequency, at frequencies less than ~0.2 Hz, the bicoherence magnitude is greater than ~0.4. Contour C is located at ~0.9178 Hz (nearest input frequency is 0.91Hz) with a self-coupling bicoherence magnitude of ~0.3737.

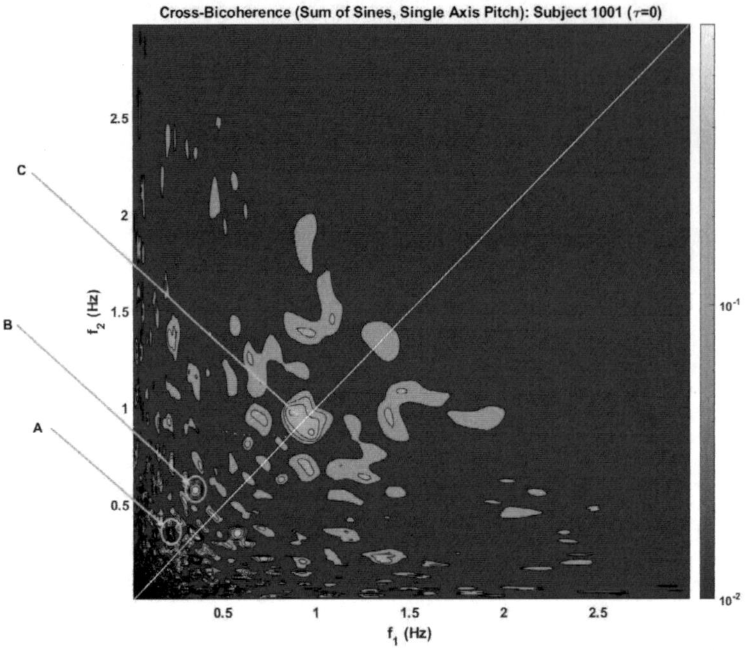

Figure 5. Cross-Bicoherence Matrix (annotated) ($\tau = 0$)

The diagonal bicoherence spectra contains several spectral peaks with significant self-coupling strength. The frequency peaks are located at A) 0.08084 Hz (b^2=0.9516), B) 0.1505 Hz (b^2=0.5750), C) 0.3424 Hz (b^2=0.5144) and 0.9178 Hz (b^2=0.3737). These close in magnitude to the following input frequencies: 0.0830 Hz, 0.1332 Hz, 0.3499 Hz and 0.910 Hz respectively. It is clear from this result that the excitation of the linear system with a multi-harmonic signal induces significant self-coupling at frequencies at or near the prescribed sum of sines input frequencies. The frequency 0.9178 Hz also has a large spectral magnitude and is very distinct (See Figure 1 (b)).

Case 2: $\tau = 0.1s$

Figure 6 (a), (b) and (c) show the cross-bicoherence matrix, the diagonal (indicated by the white dashed line) and summed bicoherence spectra respectively. As before, to make the matrix easy to read and inspect, an annotated cross-bicoherence matrix is again provided in Figure 7.

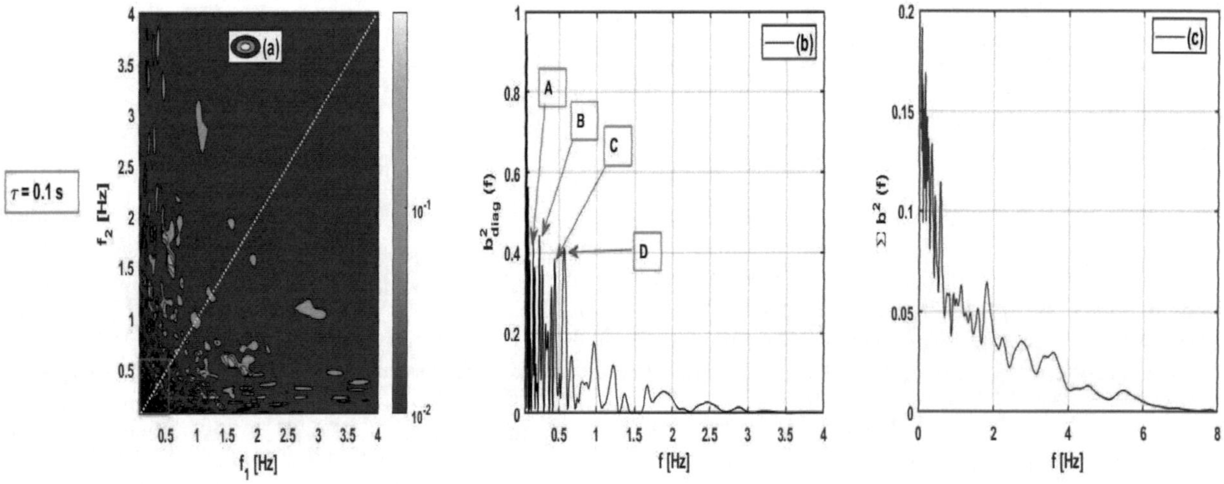

Figure 6. Cross-bicoherence matrix and spectra ($\tau = 0.1s$)

A visual inspection the matrix reveals that the there is an increase quadratic phase self-coupling along the main diagonal due to the input delay. In the off-diagonal region of the matrix, we can observe that the contours along the axes are elongated and closely space compared to the bicoherence matrix in Figure 5. The elongated contours suggest that there is phase-amplitude coupling present. In Figure 7, there are three matrix spectral features indicated, all off-diagonal. Inside the pink square, the reader will notice that the contour structure is closely spaced, harmonic and has a bicoherence magnitude greater than ~0.8. At Contour A $[b^2(0.3424\text{ Hz}, 0.5935\text{ Hz}) = 0.8253]$; this frequency pair has significant nonlinear coupling. The first entry in the frequency pair, 0.3424 Hz is closest to the input frequency 0.3499 Hz and the second, 0.5935 Hz is slightly greater than the input frequency 0.5660 Hz. At Contour B $[b^2(0.2121\text{ Hz}, 0.3523\text{ Hz}) = 0.748]$, the frequency pair is also strongly coupled and are close to the input frequencies 0.2166 Hz and 0.3499 Hz respectively. At Contour C $[b^2(0.17\text{ Hz}, 0.2647\text{ Hz}) = 0.6755]$, the frequency pair are closest to the inter-harmonics of the input frequencies, 0.1332 Hz, 0.2166 Hz and 0.3499 Hz. Specifically, the frequency 0.17 Hz is closest to ~0.1749 Hz (which is half way between 0.1332 Hz and 0.2166 Hz) and 0.2647 Hz is closest to ~0.2833 Hz (which is half way between 0.2166 Hz and 0.3499 Hz).

The diagonal bicoherence (Figure 6 (b)) shows that there is a densely packed harmonic structure (Figure 6 (a) inside the pink rectangle). There are several peaks of nearly equal bicoherence magnitude located between ~0.05 Hz and 0.5 Hz. Specifically, the peaks are located at A) 0.1571 Hz (b^2=0.4184), B) 0.2361 Hz (b^2=0.4351), C) 0.4394 Hz (b^2=0.3842) and D) 0.5646 Hz (b^2=0.4125). Above ~ 0.5Hz, the nonlinear coupling strength falls of very sharply (i.e., $b^2 < 0.2$). This spectral feature of nearly equal bicoherence magnitudes is not observed in Figure 4 (b) for the zero-time delay case. These frequencies are again near or at the prescribed input frequencies. The presence of the time delay causes the region of harmonic bicoherence to become more populated with nonlinear frequencies.

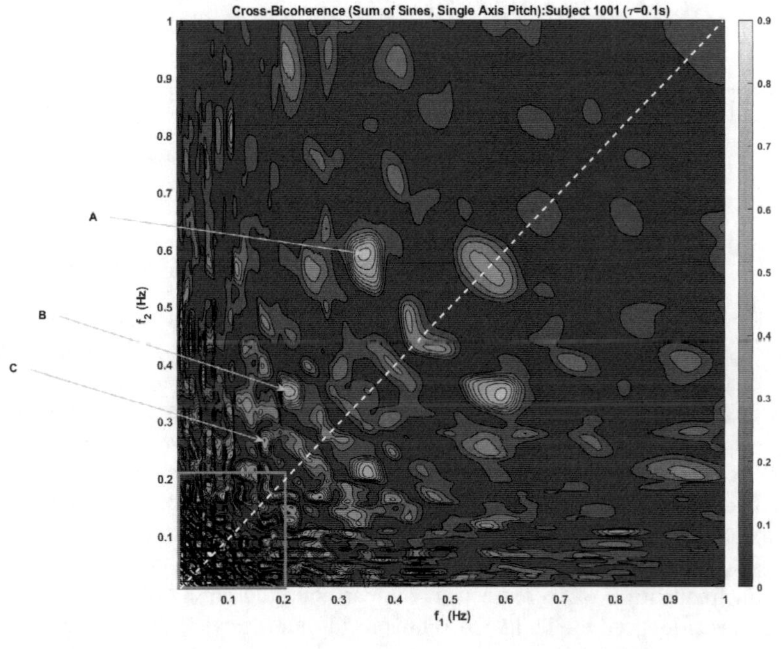

Figure 7. Cross-Bicoherence Matrix (annotated) ($\tau = 0.1s$)

On the topic of summed bicoherence, using Figures 4 (c) and 6 (c), the most distinct peaks were joined together to form two envelopes for comparison (Figure 8). In general, the summed bicoherence index tells us about the global distribution of quadratic coupling strength over the entire frequency range. In addition, it can also be used to detect any global changes to the bicoherence matrix when there are parameter variations. In Figure 8, the red line indicates the upper boundary (~1Hz) of the Rigid rotorcraft pilot coupling bandwidth prescribed by Pavel (Ref.3, see Figure A1 in Appendix). At low frequency, we can observe that both cases show similar magnitudes of coupling strength between 0.05 Hz and 0.35 Hz. Beyond 0.35 Hz, the envelopes begin to diverge. As the frequency increases, the bicoherence magnitude of the time-delayed case increases. One possible interpretation of this result is that as the pilot works harder to track the *pipper*, the bicoherence magnitude increases with the increasing frequency of longitudinal cyclic displacements. It might be instructive to examine whether similar behaviour is observed when there is a true nonlinearity present (i.e., like a rate limited actuator).

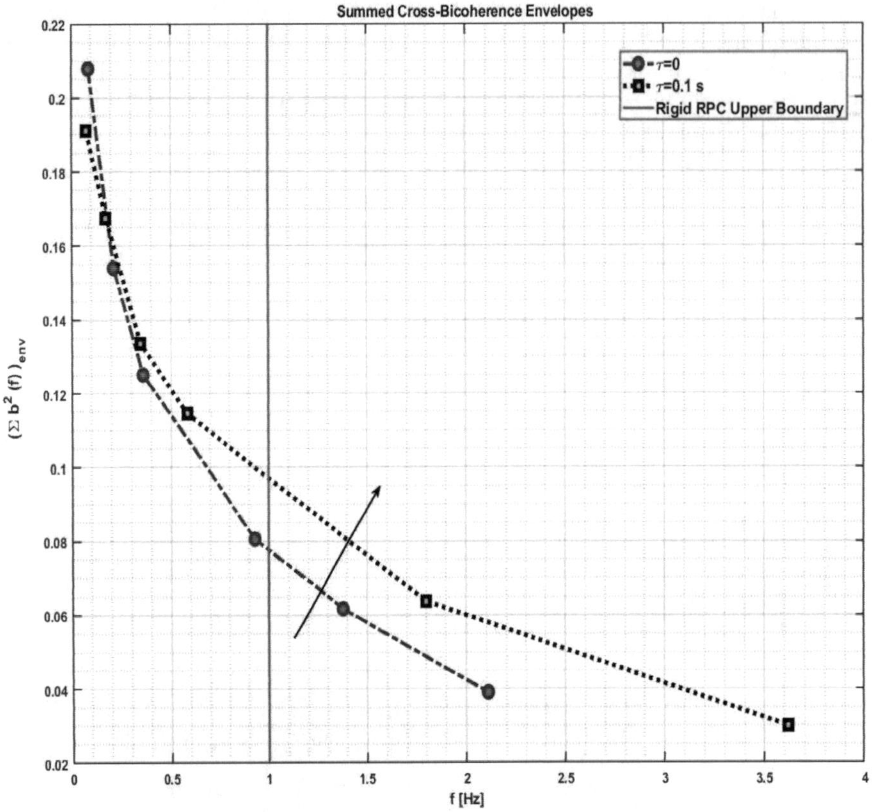

Figure 8. A comparison of summed bicoherence envelopes

CONCLUSIONS AND FUTURE RESEARCH

In this paper we have demonstrated the application of higher order spectral analysis to rotorcraft flight dynamics. A linear time invariant system was subjected to a multi-harmonic input signal. It was found that nonlinear frequency interactions could be detected and quantified by their coupling strengths. As a general observation, it was observed that the linear dynamics induces nonlinear coupling at or near the prescribed pilot input frequencies; this result is common to both cases with and without time delay. The time delay induces *quadratic phase self-coupling* (QPSC) at or near the pilot input frequencies; beyond ~0.5 Hz, the nonlinear coupling drops sharply and falls significantly below the threshold bicoherence of ~0.2. The use of a bicoherence index was also demonstrated. Specifically, the summed bicoherence index spectra is used to detect global changes in the cross-bicoherence matrix. The observed result indicates that as the pilot works to track or follow the pipper with the time delay present, the nonlinear coupling strength increases with frequency. An increase in nonlinear coupling strength can indicate potential difficulty in controlling the aircraft (as evidenced by the large excursions in pitch) as well as an increase in pilot workload. More analysis is required to confirm the initial results obtained in this preliminary study. It was pointed out in the introduction that higher order spectral analysis is most useful as a diagnostic tool especially when it is applied to a known dynamic instability. As a follow-on study, higher order spectral analysis will also be applied to fleet level lateral PIO data provided by NAVAIR. By applying this advanced signal processing method, we hope to gain some insight into the nonlinear frequency interactions involved in pilot induced oscillation.

APPENDIX

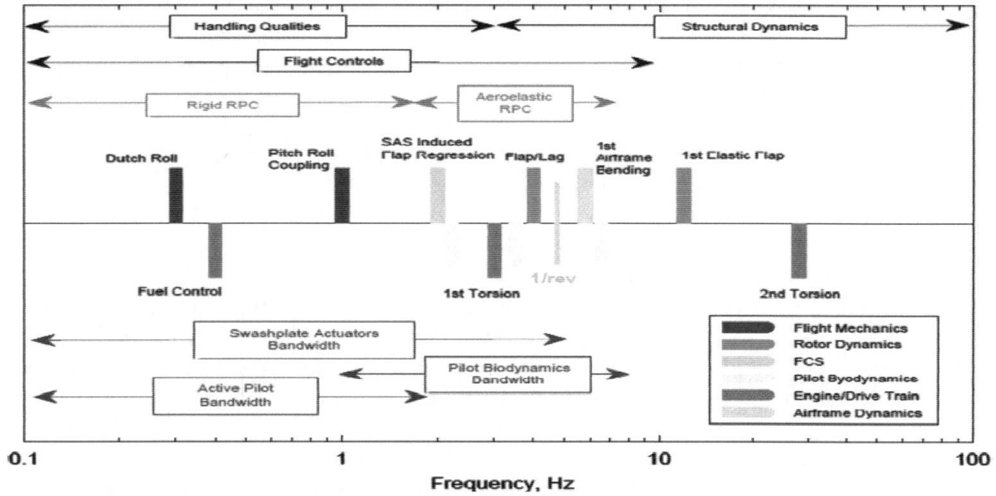

Figure A1. Modes of interest of a medium sized helicopter (Ref. 3)

ACKNOWLEDGMENTS

The authors would like to thank Matthew Rhinehart (NAVAIR) and Dr. Karen Feigh and the Georgia Tech Vertical Lift Research Center of Excellence for their support of this work.

REFERENCES

1. Baillie, S.W., and Morgan, J.M., “The Impact of Vertical Axis Characteristics on Helicopter Handling Qualities”, Aeronautical Report LR-619 NRC No. 28133, Ottawa, August 1987

2. Smith, J.W., and Edwards, J.W., “Design of a Nonlinear Adaptive Filter for Suppression of Shuttle Pilot -Induced Oscillation Tendencies”, NASA TM-813349, 1980

3. Pavel, M.D., Jump, M., Dang-Vu, B., Masarati, P., Gennaretti, M., Ionita, A., Zaichik, L., Smaili, H., Quaranta, G., Yilmaz, D. and Jones, M., 2013. Adverse rotorcraft pilot couplings—past, present and future challenges. Progress in Aerospace Sciences, 62, pp.1-51.

4. Chabalko, Christopher C. "Identification of transient nonlinear aeroelastic phenomena." PhD diss., Virginia Tech, 2007.

5. Silva W., and Dunn S., “Higher-order spectral analysis of F-18 flight flutter data”46th AIAA/ASME/ASCE/AHS/ASC Structures, Structural Dynamics and Materials Conference April 18 ,2005

6. Davis W, Lovato J, Pezeshki C., “Nonlinear spectral characterization of frequency-modulated control applied to a static airfoil shear layer”. Fluid Dynamics Conference 1994

7. Alstrom RB, Moreau S, Marzocca P, Bollt E., “Nonlinear characterization of a Rossler system under periodic closed-loop control via time-frequency and bispectral analysis”, Mechanical Systems and Signal Processing. 2018

8. Li, D., X. Li, S. Hagihira, and J. W. Sleigh. "Cross-frequency coupling during isoflurane anaesthesia as revealed by electroencephalographic harmonic wavelet bicoherence." British journal of anaesthesia 110, no. 3 (2013): 409-419.

9. Klyde, D.H., Ruckel, P., Pitoniak, S.P., Schulze, P.C., Rigsby, J., Xin, H., Brewer, R., Horn, J., Fegely, C.E., Conway, F. and Ott, C.R., “Piloted simulation evaluation of tracking MTEs for the assessment of high-speed handling qualities”, Annual Forum Proceedings-AHS International (Vol. 2018). 2018

Rotorcraft Vertical Rate Command's Impact on Workload and Handling Qualities and its Implementation-Practicality

Ian Willebeek-LeMair
Aerospace Engineer
NAWCAD
Patuxent River, MD, USA

Matthew Rhinehart
Aeromechanics Senior Engineer
NAWCAD
Patuxent River, MD, USA

ABSTRACT

An advanced Vertical Rate Command (VRC) flight control augmentation is prototyped and applied to the CH-53K™ King Stallion to investigate its impact on workload and handling qualities and to further the implementation-practicality vertical augmentation for fly-by-wire rotorcraft. The Vertical Response Augmentation for CH-53K™ (VRAK) prototype applies a model following controller to effect a VRC response type and includes a split force-displacement (SFD) control method to address challenges to the practical, rather than theoretical, implementation of rotorcraft vertical augmentation. The impact of VRC on workload and handling qualities was evaluated in formal piloted evaluations conducted in the CH-53K™ flight simulator at NAS Patuxent River using a sequence of six maneuvers flown by US Marine Corps test pilots. Results demonstrate that VRC delivers substantial benefits to workload and handling qualities in the majority of the maneuvers evaluated. Observations from independent engineering tests show that SFD enables practical application of vertical augmentation. Altogether, VRAK is demonstrated to be a logical next step in fly-by-wire rotorcraft vertical control.

NOTATION

Symbols

a_{zalt}, v_{zalt}	Rotorcraft vertical acceleration (ft/s^2) and vertical rate (fpm or ft/s)
A, B	State-space A and B matrices
F_P	Total pilot force on both collectives (lbf)
$\dot{h}$	Climb rate (fpm)
K	Arbitrary constant gain
r^2	Coefficient of determination
T	First order time constant (s)
u_h, Δu_h	Collective control signal and collective control signal increment (inches of coll.)
$UCD(s)$	Unitary collective dynamic transfer function
δ_c	Collective deflection from trim (inches of coll.)
τ	Time delay (s)
ζ	Damping ratio

Scripts

C-AIS	Collective position control signal
C-AIS CMD	Collective position command signal
CM	Command map
FPI	Forward path inversion
FB, FDBK	Feedback
$\dot{h}_{eq}$	ADS-33E vertical rate response criteria
ISM	Input signal management (from EGI)
Lead	Lead control signal
Lead CMD	Lead control signal, pre-limiting
RM	Reference model
TM	Trim map
Σ	Control signal summation

INTRODUCTION

To enhance the safety and effectiveness of the military rotorcraft mission in demanding and high-workload tasks, the Naval Air Warfare Center Aircraft Division (NAWCAD) has developed a prototype vertical control augmentation called Vertical Response Augmentation for CH-53K™ (VRAK) and used it to evaluate the workload and handling quality impacts of integrating a vertical rate command (VRC) response type on an advanced fly-by-wire rotorcraft and to investigate the technology's implementation-practicality for development and production rotorcraft. This work was performed as part of NAWCAD's Advanced Flight Control Augmentation for Pilot Workload Alleviation research effort. The premise behind this work is that control augmentation, tailored to the needs and interests of military rotorcraft and military mission elements, presents an opportunity to improve safety and enhance mission capability.

Presented at the VFS Handling Qualities Technical Meeting, Huntsville, Alabama, USA Feb. 7-8 2023.

Current state-of-the-art control systems implemented on Navy fly-by-wire rotorcraft, such as the CH-53K™, lack advanced pilot in the loop vertical axis augmentation. This research effort seeks to address that gap. This project examined the hypothesis that implementing a VRC response type on a fly-by-wire rotorcraft alongside translational augmentation would stabilize the rotorcraft's vertical channel, promote repeatability and consistency across all flight conditions, and enable increased precision and aggression leading to reduced workload and improved handling qualities. The CH-53K™ and future Navy programs such as Future Vertical Lift - Capability Set 2 (FVL-CS2) Maritime Strike and the V-22 mid-life upgrade stand to benefit substantially from the results of this research.

Figure 1. CH-53K™ King Stallion

This project selected the CH-53K™, shown in Figure 1, as its target platform. As the Navy's first full authority fly-by-wire rotorcraft, the CH-53K™'s architecture enables advanced augmentation, making this rotorcraft an ideal opportunity for this research effort. Early in the CH-53K™'s development, based on criteria from ADS-33E (Ref. 1), requirements for heavy translational axis control augmentation were put in place to promote good handling qualities and low workload. Vertical axis control augmentation offered similar benefits, but presented unique practical implementation challenges. These included: (1) the collective has immense control power, permitting the possibility of dangerous mode transition responses (MTRs); (2) the collective is central to safety, changing the collective had the potential to compromise pilots' ability to rapidly climb and autorotate; and (3) the collective is nominally non-unique trim, complicating detent management between the baseline stick-to-head (S2H) system and traditionally unique trim augmentations (like VRC). These challenges required further investigation to determine whether vertical augmentation through the collective was at all practical for development and production rotorcraft. As translational augmentation was found to sufficiently reduce overall workload, full vertical augmentation was not pursued. Even with these challenges, the dramatic handling quality and workload benefits of the CH-53K™'s translational axis augmentation strongly indicate that similar vertical axis augmentation could offer substantial benefits to both current and future Navy fly-by-wire rotorcraft.

The VRAK prototype CLAW is a modified explicit model following (EMF) control law (CLAW) that commands vertical rate proportional to pilot collective force. VRAK was developed based on the Navy's Controls Analysis and Simulation Test Loop Environment (CASTLE) model of the CH-53K™. The control system is the synthesis of CASTLE data and existing CH-53K™ control architecture. The architecture of these CLAWs is based on other state-of-the-art vertical response augmentations (Refs. 2, 3) and includes a modification called a split force-displacement (SFD) control method. This modification was designed by NAWCAD engineers and it constitutes a path forward towards practical development and production implementations of vertical augmentation for non-unique trim collective configurations. The solution addresses the implementation challenges of vertical augmentation and includes further benefits. The result is distinct from other recent research involving vertical axis augmentation which implements VRC by applying unique trim to collective position (Refs. 4, 5). With SFD, VRAK commands vertical rate according to force, rather than collective position. Collective position retains a steady-state stick-to-head link to collective blade pitch angle without forfeiting VRC. The combination of these two effects address the three challenges to implementation-practicality and enables both pilot selectability and envelope cuing while effecting a VRC response type.

A pilot evaluation of the VRAK implementation of VRC was conducted in the Containerized CH-53K™ Simulator at NAS Pax River. The event was flown with three Navy test pilots. This experiment targeted the project's primary objective: to evaluate the workload and handling quality impacts of VRC. The tasks flown were selected to cover the range of precision and aggression demanded by the military rotorcraft mission. These tasks were: (1) pattern lap, (2) hover mission task element (MTE), (3) vertical maneuver MTE, (4) aerial refueling, (5) autorotation, (6) terrain following (TERF) flight.

This paper presents its research objective and then a short, functional description of pertinent characteristics of the CH-53K™ CASTLE and simulation model. Next follows a discussion of the development and implementation of the VRAK prototype CLAW. Results informing VRC's impact on workload and handling qualities, derived from the formal pilot evaluations, and results related to VRC's implementation-practicality are presented and discussed. Implication results of SFD for envelope cueing are identified. Finally, the authors present overall conclusions.

OBJECTIVE

This research effort had two objectives: (1) to answer the question "What are the workload and handling quality impacts of implementing a vertical rate command (VRC) response type?" and (2) to identify a practical path forward towards an implementation version of this technology.

The CLAW designed for this project, the VRAK prototype, was tailored to those objectives. Two primary design drivers for VRAK were to: (1) deliver a tunable VRC response type and (2) overcome known challenges to further vertical augmentation's implementation-practicality.

The first design driver entailed developing a control architecture that would effect a vertical-rate-command zero-vertical-rate-hold response type that would comply with relevant design criteria from ADS-33E-PRF (Ref. 6), and would enable the designers to tune pilot facing features of the control system such as effective pilot bandwidth, control sensitivity, and responsiveness. The tunable, ADS-33E compliant VRC response type was achieved through the implementation of VRAK's modified EMF architecture which included a tunable reference system and an inverse plant model. The reference system and inverse plant model are discussed in the Reference System and Forward Path Inversion subsections of VRAK Architecture, respectively.

The second design driver entailed addressing three primary implementation challenges to applying vertical response augmentation to the collective in the interest of demonstrating VRC's practicality. Here, the term implementation-practicality, as applied to VRC, refers to the likelihood that implementing the technology on a development or production rotorcraft would be successful and beneficial in real, rather than theoretical, circumstances. VRAK addresses these challenges though the combination of the SFD control method modification, force cutoffs, and a collective trim map. Details on the implementation challenges and SFD are presented in the Implementation Challenges and SFD Control Method sections of this paper, respectively. It is worth emphasizing: SFD is not yet a fully mature solution and there are other strong candidates for vertical augmentation whose practicality should be the subject of future research.

IMPLEMENTATION CHALLENGES

The potential benefits of a VRC-type augmentation have long been understood by the rotary-wing handling quality community. Several implementation challenges, however, have prevented the community from realizing control systems that effect true, full-time advanced response types on the collective in production systems. This effort considers the three implementation challenges:

1. Accommodating mode transition responses (MTRs)
2. Preserving full control power
3. Managing the collective's non-unique trim

These implementation challenges arise from the fact that standard approaches to VRC augmentation replace the one-to-one correspondence between collective position and collective blade pitch with a one to one correspondence between collective position and vertical rate. Transitioning the meaning of collective position between collective blade pitch and vertical rate could produce objectionable or even dangerous mode transition responses, can limit available control power and complicate autorotation, and needs to accommodate changes in trim.

MTRs are responses developed due to transitions between control modes. MTRs may have zero or nonzero value in steady state. If a MTR has zero value in steady state, it is a mode transition transient (MTT). The MTR associated with the collective is of particular concern since it is not transient—pilots would need to arrest the MTR manually. Furthermore, modern rotorcraft typically have large configuration and operational envelopes. For these systems, transitioning between control modes where the position of the collective commands different responses has the potential to generate vertical rate responses on the order of thousands of feet per minute. These responses would be most objectionable in circumstances including landing, low-altitude flight, and TERF flight where an undesired vertical response could lead to a mishap. Accordingly, a practical implementation of VRC needs to include some mitigation of MTRs.

In the interest of safety and capability, vertical augmentation should not limit pilots' vertical control power. That is, pilots need to be able to pull full collective and autorotate should the need arise. Accordingly, any practical vertical augmentation scheme that remaps the meaning of collective position to a quantity other than collective blade pitch, e.g. vertical rate, needs to design some mechanism through which pilots can retain full vertical control power.

The fact that the collective is non-unique trim poses a challenge unique from the other inceptor axes on fly-by-wire rotorcraft. Upon entering and exiting vertical augmentation, the collective's detent needs to be re-referenced. When entering augmentation, if the vertical augmentation remaps the meaning of collective position to a one-to-one correspondence with vertical rate, then the detent needs to move to whatever position is designed to correspond to the trim vertical rate. Typically, this is chosen as the center of the collective's throw (Refs. 4, 5). Moving the detent causes forces that, in turn, move the collective. If the pilot holds the stick in place so that it does not move, they would feel the motion of the detent as a change in force-feel. These issues associated with engaging augmentation are tractable since designers are able to choose the detent's location while augmentation is engaged. Disengaging augmentation is different. Upon exiting vertical augmentation, the detent needs to be re-referenced to a specific setting: the collective blade pitch corresponding to whatever vertical rate was trimmed prior to disengage. The key issue is identifying the value of the collective blade pitch that actually corresponds to the augmentation's trim vertical rate based on limited measurements of the aircraft state and flight condition. Accordingly, a practical VRC design should accommodate for collective re-trim on mode transition.

AIRCRAFT MODEL

The rotorcraft model used to develop and evaluate VRAK was the Navy's CASTLE model of the CH-53K™. CASTLE is an aircraft simulation environment that supports desktop and flight simulator modeling of many Navy aircraft. The system functions by integrating high fidelity aircraft models. The models are based on experimental data.

VRAK was developed in a desktop design environment comprised of an interface between CASTLE and Simulink. CASTLE performed all simulation except for the CLAWs, which were handled by Simulink.

The major limitation of the CASTLE model of the CH-53K™ for this project was that it lacked a model of the aircraft's collective active inceptor system (C-AIS). The research team, therefore, developed and implemented their own model of the C-AIS in Simulink. In the desktop design environment, the C-AIS model ran in parallel with the Simulink model of the CH-53K™ CLAWs.

BASELINE CH-53K™ CONTROL LAWS

The Simulink model used in this research used a developmental version of the aircraft's CLAWs, which was deemed sufficient to support the research effort.

There are three translational control modes on the cyclic: the primary flight control system (PFCS), the automatic flight control system (AFCS), and position hold override (PHO). The response types for these modes are shown in Table 1.

The CH-53K™'s vertical axis is, by comparison, mostly unaugmented. The vertical channel supports two hold modes: altitude hold and vertical speed hold. When out of the loop, pilots can engage vertical holds regardless of their translational control mode. These hold modes, however, suspend whenever the pilot moves the collective out of detent. When the pilot is hands-on the collective inceptor—in the loop—they have exactly one option for how to fly the vertical channel, a filtered but unaugmented baseline (BL) control mode with a S2H response type.

It is informative to draw an analogy between the CH-53K™'s translational modes and its vertical modes, including the prototype VRAK mode. In this analogy PFCS is closest to BL, AFCS is closest to VRAK, and PHO is closest to the CH-53K™'s vertical hold modes. This analogy underlines a design intent of VRAK. VRAK is intended to be an intermediate level of augmentation between the baseline S2H response type and current vertical hold modes. (I.e. it does not replace the vertical holds.)

VRAK ARCHITECTURE

VRAK is a limited-envelope, modified EMF prototype control law for the CH-53K™. As shown in Figure 2, VRAK consists of seven subsystems: (1) command map (CM), (2) reference system (RS), (3) collective trim map (TM), (4) forward path inversion (FPI), (5) feedback (FB, FDBK), (6) limiting, and (7) the collective active inceptor system (C-AIS). The other subsystems shown in Figure 2 represent the CH-53K™'s primary flight control partition (PFCP), the bare airframe, and the aircraft's sensors. Specifically, VRAK relies on vertical motion measurements from the CH-53K™'s embedded GPS inertial navigation systems (EGIs). The following subsections detail the design of each VRAK subsystem.

Command Map

The command map (CM) transforms measured force on collective, F_P, into a commanded vertical rate, $v_{\text{zalt}}^{\text{CM}}$. Figure 3 shows VRAK's current CM. It has not yet been tuned with pilot feedback. This map features a breakout, a linear region where commanded vertical rate is proportional to applied force, and force cutoffs. If the pilot pushes or pulls more than the cutoff threshold, VRAK is disengaged. This cutoff promotes safe autorotations and guarantees that the pilot is always able to pull full power.

Currently, VRAK does not support re-trimming due to limits in schedule. Trim is fixed at zero vertical rate. A production version of VRAK would need to consider adding the capability to trim to different vertical rates.

Reference System

The reference system (RS) is comprised of two subsystems: a reference model (RM) and a reference governor (RG). The reference model follows the second order transfer function

$$v_{\text{zalt}}^{\text{RM}} = \left(\frac{{\omega_n^{\text{RM}}}^2}{s^2 + 2\zeta^{\text{RM}}\omega_n^{\text{RM}} s + {\omega_n^{\text{RM}}}^2} \right) v_{\text{zalt}}^{\text{CM}} \quad (1)$$

where $v_{\text{zalt}}^{\text{CM}}$ is the commanded vertical rate from the command map, ω_n^{RM} is the natural frequency of the reference model, ζ^{RM} is the damping ratio of the reference model, s is the Laplace domain variable, and $v_{\text{zalt}}^{\text{RM}}$ is the reference vertical rate. ω_n^{RM} and ζ^{RM} are set to 2.1 rad/s and 1.0, respectively.

Table 1. Baseline CH-53K translational control modes

TCM		Longitudinal Cyclic	Lateral Cyclic	Pedals
PFCS		Rate command / attitude hold	Rate command / attitude hold	Rate command / heading hold
AFCS	(low speed)	Attitude command / groundspeed hold	Attitude command / groundspeed hold	Rate command / heading hold
AFCS	(high speed)	Attitude command / airspeed hold	Attitude command / attitude hold	Rate command / heading hold
PHO	(low speed only)	Translational rate cmd. / position hold	Translational rate cmd. / position hold	Rate command / heading hold

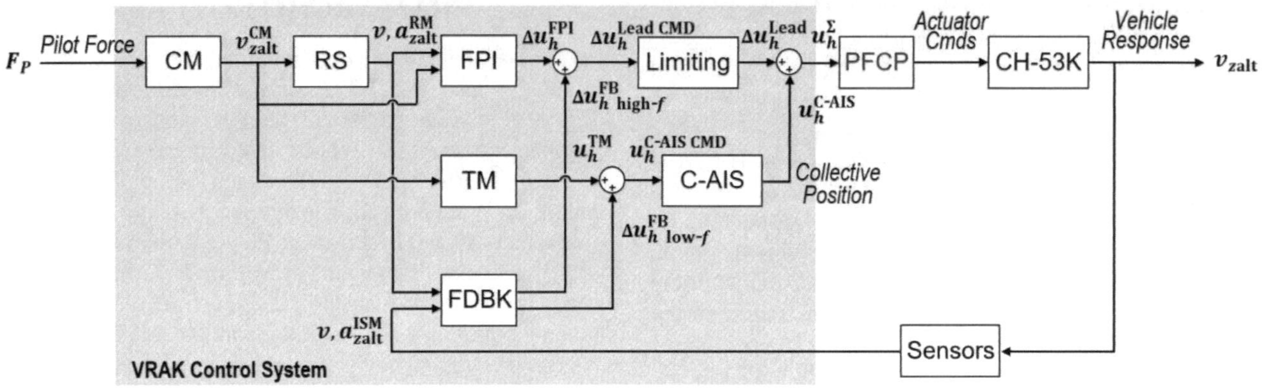

Figure 2. VRAK Signal Flow Diagram

VRAK's reference governor follows a standard architecture and is configured to limit the lead increment from the FPI subsystem, $\Delta u_h^{\mathrm{FPI}}$, to a maximum magnitude of 1.0 inch of rotor head. The purpose of the reference governor is to prevent feedback windup in the position of the collective. If it were not governed, the windup would present itself to pilots as undesirable feel characteristics.

Trim Map

The trim map (TM) converts the commanded steady vertical rate, $v_{\mathrm{zalt}}^{\mathrm{CM}}$, into a commanded inceptor position, u_h^{TM}. This conversion is scheduled on altitude and groundspeed. The VRAK prototype's trim map is limited-envelope. Its limits are visualized in Figure 4. The trim map was developed using data collected from the Navy's CASTLE model of the CH-53K™. For the VRAK prototype, the domain of this data was constrained to a subset of the CH-53K™'s flight envelope.

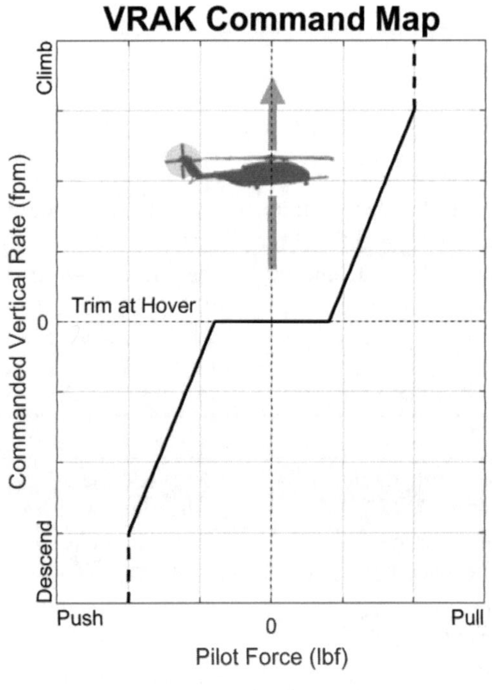

Figure 3. VRAK Command Map

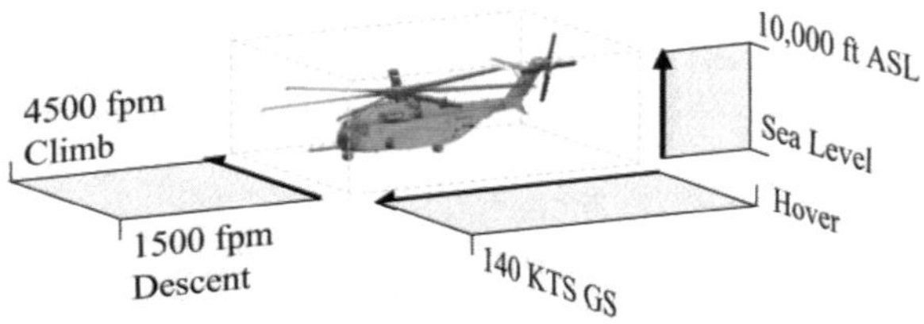

Figure 4. VRAK prototype's trim map limits

Forward Path Inversion

The forward path inversion (FPI) subsystem implements a first order inverse plant model to develop a feedforward control signal used to add lead to the rotor head, increasing the system's effective pilot bandwidth. Other EMF architectures use similar low order and zeroth-order over first-order transfer function fits for their vertical inverse models (Refs. 4, 5). VRAK's FPI implements

$$\Delta u_h^{\mathrm{FPI}} = B^{-1}\left(a_{\mathrm{zalt}}^{\mathrm{RM}} - A\left(v_{\mathrm{zalt}}^{\mathrm{RM}} - UCD(s)\, v_{\mathrm{zalt}}^{\mathrm{CM}}\right)\right) \quad (2)$$

where $\Delta u_h^{\mathrm{FPI}}$ is the FPI lead control signal, $a_{\mathrm{zalt}}^{\mathrm{RM}}$ is the RM vertical acceleration, $v_{\mathrm{zalt}}^{\mathrm{RM}}$ is the RM vertical rate, $v_{\mathrm{zalt}}^{\mathrm{CM}}$ is the commanded vertical rate, A, B (scalar) are the A and B matrices from the first order model of vertical rate, and $UCD(s)$ is a unitary collective dynamic model. The terms A and B were calculated using the CASTLE model of the CH-53K™ flying in hover at 5,000 ft altitude. One linearization point was determined sufficient for the VRAK prototype, but a production implementation would schedule these parameters on flight condition. The factor $UCD(s)$ is a second order transfer function model of the dynamics of the collective inceptor in the CH-53K™ simulator that has a unitary DC gain.

The major difference between FPI and other inverse-plant-model feedforward approaches to EMF (Refs. 2, 3, 4, 5) is that FPI does not produce the complete control signal necessary to follow the reference trajectory. (This is part of VRAK's SFD control method.) FPI is, instead, designed to produce only a lead signal. This lead is added to the direct S2H-effect of the position of the collective. The difference between VRAK and other EMF approaches appears in the term $UCD(s)\, v_{\mathrm{zalt}}^{\mathrm{CM}}$. Including this term renders the steady-state value of Δu_h^{FPI} zero. The term $UCD(s)\, v_{\mathrm{zalt}}^{\mathrm{CM}}$ converts FPI from a standard inverse-plant feedforward controller into a system that produces a lead signal. If this term were not present, FPI would effectively double the steady control signal already provided by the collective itself. This approach to converting FPI from a complete inverse-plant-model feedforward controller into one that only produces lead could likely be improved and should be studied in future work.

The reason that standard approaches to EMF control do not attempt to convert their feedforward path into lead-only systems is that they choose to accomplish all control through the EMF control system. In this choice, they eliminate the S2H-effect of their inceptor. VRAK, in contrast, intentionally maintains the S2H-effect of the collective. As will be discussed in the SFD Control Method section, maintaining the S2H-effect of the collective enables the control system to match the steady state positions of the collective between the VRAK and BL vertical control modes—a key enabler implementation-practicality.

Feedback

VRAK directly uses the vertical axis feedback system from the CH-53K™'s baseline CLAWs. The system feeds back a linear combination of acceleration error and velocity error and integrated velocity error scheduled on flight condition.

The baseline system splits its feedback signal into a high- and low-frequency component. This split is effected with a combination of a washout filter and rate limiting. After subsequent limiting, the high-frequency component is applied directly to the rotor head. The low-frequency signal is fed back through the collective by driving the position of the detent.

VRAK uses the same high- and low-frequency feedback signals from the feedback system. The high frequency feedback, $\Delta u_{h\ \text{high-}f}^{\text{FB}}$, is added to the lead increment from FPI, Δu_h^{FPI}, forming the total commanded lead control increment $\Delta u_h^{\text{Lead CMD}}$. After limiting, this lead control signal is applied directly to the rotor head. The low frequency feedback, $\Delta u_{h\ \text{low-}f}^{\text{FB}}$, is added to the commanded inceptor position, u_h^{TM}. This total position command for the C-AIS, $u_h^{\text{C-AIS CMD}}$, is used to actively drive the position of the collective in the pilots hand.

Similar to other EMF control systems, VRAK's stability properties are fixed by its feedback system. The fact that VRAK relies on the feedback system native to the CH-53K™ CLAWs, therefore, means that VRAK's stability properties are identical to that of the baseline vertical system and do not require additional broken-loop stability testing beyond those already flown.

Limiting

The limiting subsystem limits $\Delta u_h^{\text{Lead CMD}}$ to a maximum magnitude of 1.5 inches of rotor head. The CH-53K™ has a total collective throw of 10 inches. The lead signal therefore has 15% authority. The limit for $\Delta u_h^{\text{Lead CMD}}$ is 0.5 inch greater than the RG's limit of 1.0 inch for Δu_h^{FPI}. The margin of 0.5 inch guarantees that the high frequency feedback, $\Delta u_{h\ \text{high-}f}^{\text{FB}}$, will always have at least 0.5 inch available, promoting robustness.

This 1.5 inch limit is the absolute maximum difference allowed between the real world position of the C-AIS and the real world main-rotor collective angle. Limiting that difference prevents the rotor head from getting too far from what is cued to the piloted via inceptor position. Collective position maintains its role as a (now gross, but still critical) proprioceptive cue to pilots on main-rotor collective angle.

Collective Active Inceptor System

Active Inceptor Systems (AISs) enable the motion of the inceptor to be scheduled independent of the force applied by the pilot. That is, AISs decouple the motion of an inceptor from the force applied to it. AISs have control laws that use measurements of applied force as a command for inceptor position and control servos to move the inceptor accordingly.

VRAK uses the CH-53K™'s C-AIS by programming its motion in response to commanded vertical rate and using the resulting position, $u_h^{\text{C-AIS}}$, as a control input. VRAK controls collective blade pitch with $u_h^{\Sigma} = u_h^{\text{C-AIS}} + \Delta u_h^{\text{Lead}}$. This control is accomplished in precisely the same way that BL controls collective blade pitch with $u_h^{\text{C-AIS}}$ alone. When VRAK's lead control signal, Δu_h^{Lead}, is zero, the position of the collective is stick-to-head. Since Δu_h^{Lead} is zero in steady-state, $u_h^{\text{C-AIS}}$ is solely responsible for steady-state collective blade pitch. Elsewhere in this paper, $u_h^{\text{C-AIS}}$'s S2H contribution to collective blade pitch is referred to as the 'S2H-effect' of collective position.

SFD CONTROL METHOD

The split force-displacement (SFD) control method is the most innovative feature of VRAK. SFD is an original concept developed by engineers at NAWCAD. SFD is intended to solve the three challenges to implementing vertical axis augmentation on military rotorcraft identified in the Implementation Challenges section of this report.

The three challenges all arose in the context of augmentations that replaced the one-to-one correspondence between collective position and collective blade pitch with a one to one correspondence between collective position and vertical rate. VRAK does not make that replacement.

VRAK changes the force response of the collective from a legacy mass-spring-damper plus friction response type to one where inceptor force commands vertical rate and inceptor position is (remains) linked to collective blade pitch angle. Both force and position are used as inputs for the control system. The control signal developed from force is chosen to be transient. This has the desirable implication of guaranteeing inceptor position retains its one-to-one correspondence with collective blade pitch in steady state. Retaining the one-to-one correspondence between position and blade pitch addresses all three implementation challenges.

The terminology "split force-displacement control method" arises from the following analogies. For a given control path, standard control architectures use inceptor position, or displacement from trim, as their only input—these systems have a displacement control method. Other systems, like the F-16 sidestick, use inceptor force as their input—these systems have a force control method. VRAK controls vertical rate using both inceptor force and inceptor position as inputs with split responsibilities—VRAK has a, so called, split force-displacement control method.

The notion of the SFD control method refers to a specific approach to managing inceptor force, inceptor displacement, and a system's controlled output that leverages the unique capabilities of active inceptor systems. SFD is not specific to VRAK. In principle, the SFD concept could be applied to any control system with an active inceptor.

With an SFD control method, inceptor force commands inceptor position and both inceptor force and position contribute to commanding a response from a controlled output. Inceptor force is intentionally chosen to have a transient-only effect on the output. As the feedback loop closes on the output, owing to the fact that force drives position (not transient), the design choice for force-to-output transience drives the inceptor position to converge to a position that causes the steady output to match the steady command.

More precisely, a control system satisfying the following criteria is defined to have a split force-displacement control method:

1. Inceptor position has a (near) one-to-one steady-state correspondence with the controlled output
2. The commanded value of the controlled output is determined by inceptor force only
3. Inceptor force controls inceptor position in a manner that includes feedback of the commanded and real-world values of the controlled output
4. Inceptor force and inceptor displacement both effect the real world controlled output
5. The effect of inceptor force on the real world controlled output is transient

In steady-state, this SFD-enabled control system drives the inceptor to rest in the (nearly) unique position that causes zero error between the commanded and real-world values of the output—for VRAK, the collective comes to rest in the position whose S2H-effect on vertical rate maintains the commanded vertical rate. This position is identical between VRAK and BL.

SFD's purpose—the reason it is included in VRAK—is to match the steady-state position of the collective between VRAK and BL. This matching is a key enabler of implementation-practicality. If a pilot flying with VRAK established some steady vertical rate and then downshifted from VRAK to BL, they would, theoretically, experience little to no mode transition response because the position of their collective has an identical impact on vertical rate between the two modes.

Note that even though VRAK replaces the legacy mass-spring-damper plus friction response type of the collective with a new design, it still preserves expected stick behavior. With VRAK engaged, the pilot can always command a return to trim by simply releasing the inceptor, just like a legacy system. Since VRAK's unique *force* trim is at zero pounds, the inceptor returns to the detent. The position of the detent, however, is not fixed. The detent will move to account for the demand of maintaining the trim vertical rate across different flight conditions. In this sense, VRAK has a non-unique *position* trim. Overall, however, from the pilot's perspective the force-feel of VRAK is not very different from legacy systems.

It should be emphasized that the SFD control method is distinct from trim-follow-up. In trim follow up, the detent converges towards whatever input is being made by the pilot. Eventually, the detent reaches the inceptor position and the pilot is able to release pressure while maintaining their input. With SFD, the active inceptor system is continuously driving the stick to move the inceptor based on any changes to the commanded inceptor position. This backdriving does not move the detent onto the inceptor position. Instead, inceptor position maintains a steady-state one-to-one link with the commanded output. For VRAK, the inceptor position is linked one-to-one with steady-state vertical rate through the nearly one-to-one relationship between steady-state vertical rate and collective blade pitch.

FLIGHT SIMULATOR INTEGRATION

Simulation Facility

VRAK was integrated into the containerized CH-53K™ flight simulator at NAS Patuxent River. The simulator includes a full recreation of a CH-53K™'s cockpit. The majority of the systems are production hardware with a few exceptions, namely the AISs. The production aircraft has AISs supplied by BAE. The inceptors in the simulator are supplied by Sterling. The Sterling AISs replicate most of the functionality of the BAE systems, but impose some limitations.

Limitations

When VRAK was implemented in the simulator, the Sterling inceptors were unable to produce the mechanical characteristics—inertia and damping—programmed by the CH-53K™. This impacted the behavior of the control system as the $UCD(s)$ model in the FPI subsystem needed to be reconfigured to match the measured dynamics of the inceptors instead of matching the programmed characteristics. This likely degraded the model following performance of the system.

When testing VRAK in the simulator, the research team observed attitude transients whenever aggressive inputs were made on the collective. These transients are likely due to the fact that the CH-53K™ uses fixed-gain crossfeed decoupling instead of a dynamic method such as a dynamic decoupling matrix (Ref. 2). A production implementation of VRC would likely benefit from a dynamic decoupling method.

ANALYSIS OF THE INTEGRATED SYSTEM

VRC Performance

Figure 5 illustrates the vertical rate and collective motion responses of the CH-53K™ flying with AFCS and VRAK to a stepped pull of force in two different flight conditions: hover and level forward flight at 80 kts both at 5,000 ft altitude. The data for these plots was generated from a test event in the CH-53K™ simulator. Figures 5.a and 5.c show the hover responses and 5.b and 5.d show the forward flight responses. For both vertical rate responses, the dashed black line shows the commanded vertical rate, the red line shows the reference vertical rate, and the blue line shows the response vertical rate. For both collective motion responses, the solid black line shows the real world motion of the collective, the blue line shows the command to the collective, and the dashed cyan and magenta lines show the collective's detent. This figure demonstrates that VRAK delivers a VRC response type with strong model following and reasonable inceptor motion.

ADS-33E Compliance

ADS-33E-PRF outlines two criteria for vertical rate response types found in sections 3.3.10.1 and 3.4.3.2 (Ref. 1). ADS-33E specifies that the vertical rate response should have a qualitative first-order appearance for at least 5 seconds following a step collective input adhering to

$$\frac{\dot{h}}{\delta_c} = \frac{Ke^{-\left(\tau_{\dot{h}_{eq}} \cdot s\right)}}{T_{\dot{h}_{eq}} s + 1} \tag{3}$$

where $\dot{h}$ is climb rate, δ_c is collective input, K is an arbitrary constant gain, $\tau_{\dot{h}_{eq}}$ is time delay, $T_{\dot{h}_{eq}}$ is the first order time constant, and s is the Laplace domain variable. In the remainder of this paper, subscripts are dropped from τ and T in the interest of legibility.

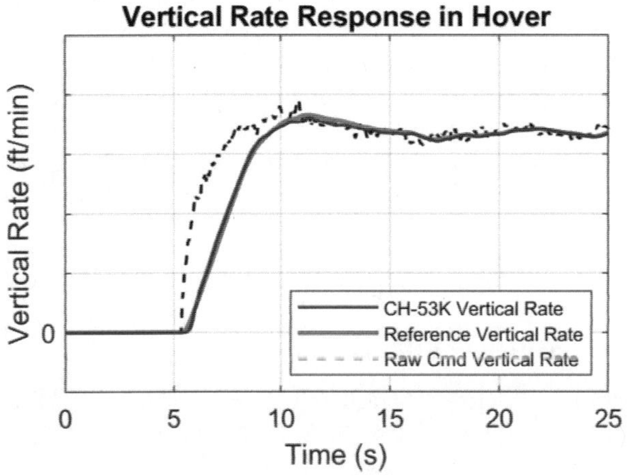

Figure 5.a Vertical Rate in Hover

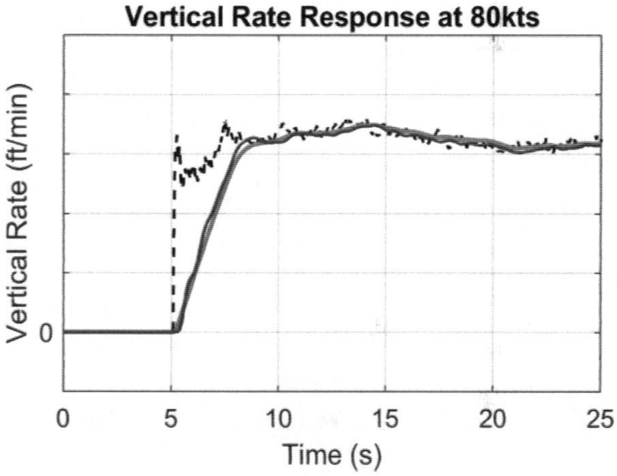

Figure 5.b Vertical Rate in Forward Flight

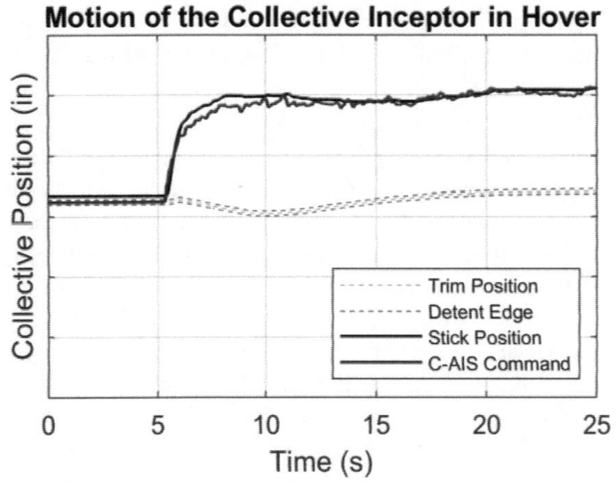

Figure 5.c Collective Motion in Hover

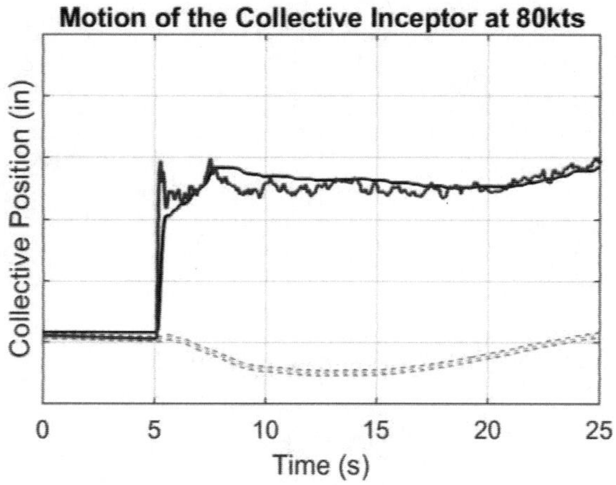

Figure 5.d Collective Motion in Forward Flight

Figure 5. Vertical Rate and Collective Motion Step Responses of the CH-53K Flying with AFCS and VRAK

ADS-33E predicts handling qualities according to Tables VII and VIII. Table VII, which specifies hover parameters, is reproduced here as Table 2. For these predictions to apply, the vertical rate response of the system must adhere to the first order response with a coefficient of determination, r^2, between 0.97 and 1.03.

Table 2. From ADS-33E: Maximum values for height response parameters – hover/low speed

LEVEL	T (sec)	τ (sec)
1	5.0	0.20
2	∞	0.30

VRAK is compliant with the ADS-33E criteria for Level 1 in both hover/low speed and forward flight. Figure 6 and Table 3 show examples for hover flight starting at 5,000 ft. In Figure 6, the normalized fit first order response, in blue, lies in the light blue region which indicates Level 1. In Table 3, VRAK achieves Level 1 in all categories and has a coefficient of determination within the prescribed tolerances. The time constant, T, for the fitted first order response is much faster than ADS-33E's criterion. The research team found that this faster time constant led to improved handling qualities. Additionally, the fitted time constant, τ, is negative. The negative value indicates there is very little delay. ADS-33E specifically states that negative values are admissible.

Table 3. VRAK's ADS-33E Compliance

	K	T	τ	r^2
Level 1	–	$T < 5$s	$\tau < 0.2$s	$r^2 \approx 1$
VRAK	1.163	2.385s	-0.021s	0.994

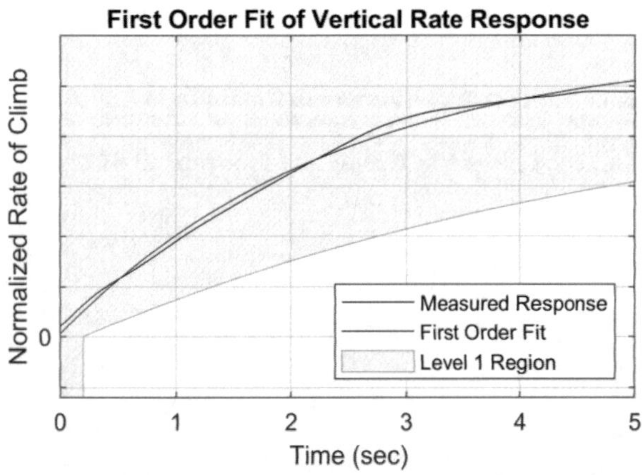

Figure 6. VRAK's ADS-33E Compliance

EVALUATION SETUP

Evaluation Overview

A formal, piloted handling quality (HQ) evaluation was performed on the VRAK realization of VRC in the CH-53K™ containerized simulator. The purpose of this evaluation was to establish the potential workload and HQ benefits of the VRC response type. The evaluation did not fully assess the behavior of VRAK concerning implementation, though comments on the topic were solicited. Three pilots participated in the formal HQ testing with all being USN Test Pilot School graduates and all with at least some experience in CH-53K™. Pilot A is a high-time CH-53E pilot with fewer than six months of experience with the CH-53K™, Pilot B has several years of experience with the CH-53K™, and Pilot C has been with the CH-53K™ since very close to the beginning of the platform's test program. The diversity of experience between these pilots strengthens the quality of the results developed by this project. The handling quality rating (HQR) scores given by these pilots (Ref. 6) and the comments they made form the basis of the major results presented in this report.

Pilot Evaluation Task Definition

Piloted evaluations included six mission tasks/mission task elements: Pattern Lap, Precision Hover MTE, Vertical Maneuver MTE, Aerial Refueling, Autorotation, and TERrain Following (TERF) flight. The first three of these events were performed in all four augmentation configurations: baseline with PFCS (BL+PFCS), baseline with AFCS (BL+AFCS), VRAK with PFCS (VRAK+PFCS), and VRAK with AFCS (VRAK+AFCS). During the last three tasks, pilots were allowed to choose which configurations they wanted to fly though they all were asked to fly a baseline case and a VRAK case for comparison. For all of the simulations, the aircraft was configured at the basic design gross weight of 60,000 lbf and a mid center of gravity.

RESULTS

Pilot Evaluation of VRC

Pattern Lap. During this task, pilots were asked to fly standard pattern laps with each of the four augmentation configurations. Pilots were directed to target and attempt to hold a specific altitude of their choice so that they could evaluate the workload difference between BL and VRAK in a non-precision/non-aggressive context. HQRs were not collected as there were no formal task definitions or performance tolerances.

All three pilots noted that VRC has only a marginal, but positive, impact on workload and handling qualities during non-precision/non-aggressive flying. These results were expected due to relatively low workload nature of the task. Just as importantly, there were no real negative comments for this task. Pilot comments indicate that there are potential benefits to be found with the addition of a pilot selectable, non-zero vertical rate trim functionality. This vertical rate trim functionality would constitute what is sometimes called a *workload relief mode* for the pilot. The research team recommends that intuitive vertical rate trim should be pursued in future work.

Precision Hover MTE. Pilots were then asked to fly the Precision Hover MTE as defined in ADS-33E (Ref. 1). The research team used the Hover MTE Cargo/Utility criteria in

ADS-33E to judge adequate and desired tolerances for this particular event, in line with the CH-53K™ nominal mission.

During their evaluations, all three pilots commented that they felt the time constraint given in ADS-33E was not appropriate for the CH-53K™, though the same tolerances were used for both sim and flight testing of the CH-53K™ during development. As they predicted, there were several flights where time was the only parameter outside of desirable. Pilot A and Pilot C gave their HQRs using the ADS-33E desired and adequate times of 5 and 8 seconds, respectively. Pilot B, however, decided to judge adequate and desired timing themselves. The HQR scores given by the pilots are listed in Table 4.

Table 4. Precision Hover MTE HQRs

Augmentation	Pilot A	Pilot B	Pilot C
BL+PFCS	HQR 5	HQR 5	HQR 7
VRAK+PFCS	HQR 5	HQR 5	HQR 7
BL+AFCS	HQR 3	HQR 3	HQR 5
VRAK+AFCS	HQR 5*	HQR 2	HQR 5**

* The only parameter outside of desired on Pilot A's flight was time at 5.8 sec. Pilot A commented "Where [I am] falling short of desired performance [has] nothing to do with the vertical axis and [has] everything to do with stabilizing longitudinally and laterally, so [I can't] fault VRAK... [comparing] AFCS and PFCS without VRAK versus AFCS and PFCS with VRAK, the workload between those two was significantly reduced as a result of VRAK."

** The only parameter outside of desired on Pilot C's flight was time at 6.1 sec. Pilot C's commented "For a task that starts at and tries to maintain zero VSI, VRAK was definitely an improvement over BL in the sense that you didn't have to hunt for zero VSI—you knew you had it."

Taking the HQRs by themselves does not provide an adequate picture of the results, so relevant pilot comments are included. Considering both, two substantive conclusions can be drawn. First, with PFCS, the high workload in the Rate Command / Attitude Hold (RCAH) cyclic mostly overshadowed any workload or handling quality benefits in the collective from VRAK. That is, the cyclic drove the HQR ratings for the PFCS cases in ways that VRAK could not help. Second, with AFCS, adding VRC through VRAK offered consistent workload reductions and handling quality improvements when compared to AFCS with BL vertical control.

The major takeaway from the Hover MTE tests is that all three test pilots saw qualitative reductions in workload and consequently improvements of handling qualities with the addition of VRC. The effects were most evident when flying with AFCS as the mediocre handling qualities of PFCS masked any benefits of VRC.

Vertical Maneuver MTE. Pilots were next asked to fly the Vertical Maneuver MTE as defined in ADS-33E (Ref. 1). There was no dedicated visual model for the Vertical Maneuver in CASTLE. A visual model for a scaffold/tower with clear features at the correct height increments was used instead. The scaffold provided good cues for the vertical tolerances but no absolute indications for lateral-longitudinal translation or heading. Pilots were asked to judge for themselves if they felt their heading, longitudinal, and lateral deviations were within tolerances. The criteria used to judge adequate and desired tolerances for this particular event were those listed under Cargo/Utility in the Vertical Maneuver MTE section of ADS-33E. The HQRs given by the pilots are listed in Table 5.

Table 5. Vertical Maneuver MTE HQRs

Augmentation	Pilot A	Pilot B	Pilot C
BL+PFCS	HQR 3	HQR 4	HQR 4
VRAK+PFCS	HQR 3	HQR 4	HQR 4
BL+AFCS	HQR 2	HQR 3	HQR 3
VRAK+AFCS	HQR 2	HQR 2	HQR 2

The story told by these ratings and the pilot comments is that VRC dramatically improves predictability in the vertical axis, which allowed aggressive control inputs while keeping workload low, making the maneuver trivial. Pilot A "really enjoyed the responsiveness of [VRC]" and felt that it made the system very controllable and predictable. Pilot B commented that "in stick-to-head you adapt your control strategy where you know you need to lead" and make corrections but VRAK is "more predictable once you get used to it." After flying the MTE with VRAK, Pilot C noted that VRC had much less drift in vertical rate than BL and predicted that they might be able to fly the vertical maneuver with zero lead, making only two inputs. In their next attempt, they proved their own prediction by flying the MTE with exactly two inputs on the collective and nothing else. There were no improvements in HQR while in PFCS as the position keeping portion of the task was still a factor even when the vertical axis was much improved by VRAK.

Figure 7 compares two of the Vertical Maneuver MTEs flown by Pilot C. The first attempt, in red, was made in BL+AFCS, without VRAK. The assigned HQR for this attempt was 3. The altitude profile, in the top sub-plot, meets all desired criteria for the MTE. Each altitude transition required several large stick inputs and some rapid tracking to set and maintain the subsequent hover. Each set of large inputs and tracking are illustrated in the second and third sub-plots. In their BL+AFCS run, Pilot C checked their descent too soon and had to apply additional down pressure to adjust. This is indicative of the generally poor predictability of the unaugmented, stick-to-head response type.

The second attempt, in blue, was made flying VRAK+AFCS. The assigned HQR for this attempt was 2 thanks to the significantly reduced workload in the vertical axis. The altitude profile, in the top plot, meets all desired criteria for the MTE.

Performing the maneuver, this time, required only two stick inputs—a simple well timed pull and push. There was no hunting necessary at the end of the maneuver as with BL. The difference in workload is reflected the inceptor position and force graphs. VRC requires much less work than BL.

In their run with VRAK, Pilot C was able to stay well within desired tolerances on all criteria and with minimal workload. Pilot C stated that they needed "zero compensation in the collective" and that "from purely counting stick inputs, [this performance] was the best possible." Pilot C did not touch the cyclic, did not touch the pedals, and made exactly two inputs on the collective. It was Pilot C's opinion that VRC enabled them to fly this particular MTE with the lowest workload desirable. The takeaway is that VRC enabled the pilot to achieve equal or better performance with fewer inputs, less stick deflection, and minimal compensation required.

Aerial Refueling. Aerial Refueling (AR) was simulated at 5,000 ft altitude and 100 kts (which is not a nominal tanking speed), flying behind a KC-130 tanker visual model. The tanker model had drogues and hoses extended on both sides though pilots mainly chose to fly to the left basket due to the right offset of the AR probe. The baskets were rigidly fixed to the tanker model and no turbulence was simulated. The pilots began their AR flights in the observation position and then flew standard AR maneuvers including approach/contact, maneuvering to refueling position, and disengagement as desired. The pilots were instructed that they could select their flight augmentation on their own and that they could transition between different levels of augmentation whenever they wanted. Though the limitations of the sim as noted above could impact the results, this task was selected to capture a precision/non-aggressive set of maneuvers in forward flight.

Pilots A and C both thought VRC was excellent for AR. Pilot A called VRC an "incredible enhancer" and Pilot C stated that they believed that "going from BL to VRAK provides the same workload reduction as going from PFCS to AFCS." In contrast, Pilot B was less enthusiastic of VRC for AR. Pilot B stated that they prefer as little augmentation as possible near the basket. That difference of preference is notable, but not necessarily surprising as pilots have diverse preferences for augmentation in different situations as predisposed by background, experience, training, natural piloting strategy, etc.

The major takeaway from the AR testing is that two of the three pilots found that VRAK's VRC response type provided substantial benefits to workload and handling qualities during AR. The third pilot's opinion was different, but concurrent with the fact that pilots have diverse preferences for augmentation near the basket. It is possible that, with additional exposure to VRAK, Pilot B would adapt for an overall better result than with BL. Regardless, this result points to a conclusion discussed later that the option for pilots to select VRAK or BL should be provided.

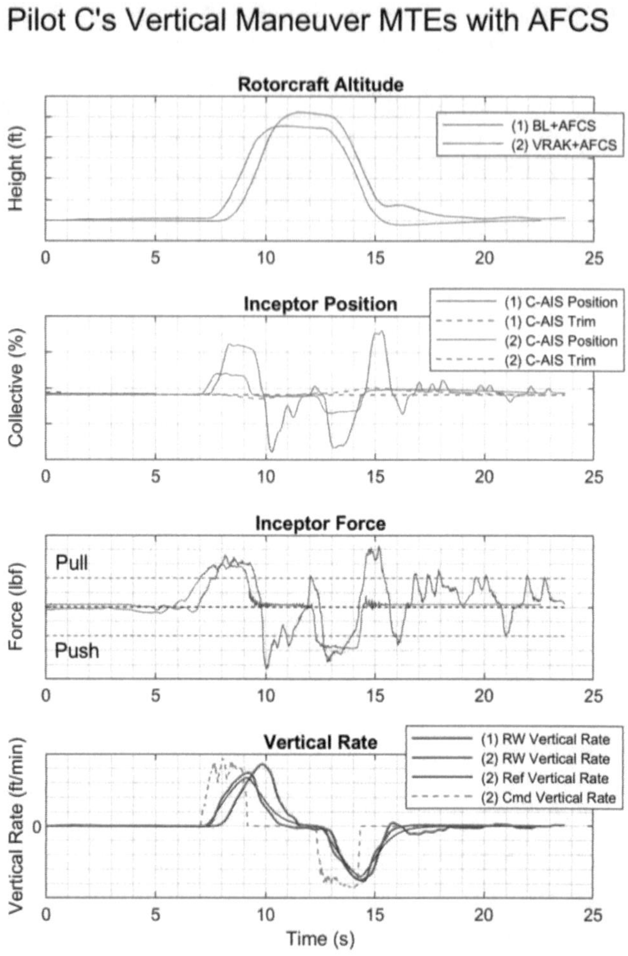

Figure 7. Comparison of Pilot C's scored BL+AFCS (red) and VRAK+AFCS (blue) flights

Autorotation. A major concern with any additional augmentation on the collective axis is how it responds to the need to enter an autorotation (either due to engine failure, tailrotor failure, or otherwise). In true emergency scenarios, there may be little warning of the impending problem and a rapid response by the pilot is typically required to preserve Nr and stabilize in a steady state descent. It follows that major question that needed to be answered was, "Will VRAK impede a pilot's efforts to enter and stabilize in an autorotative state?" The results of the sim test indicate "no", VRAK poses no significant barrier to entering an autorotative state. In autorotation evaluations, pilots executed standard power-recovery autorotations from 5,000 ft at 100 kts using both VRAK+PFCS and VRAK+AFCS. All pilots were able to successfully enter, stabilize in, and terminate to powered flight autorotations with reasonable workload. Pilot A noted that performing a full autorotation with VRAK engaged was possible, but "slightly more difficult than with BL" and that the "hardest part was finding the collective input to achieve the target Nr." The pilots found that the lowest workload option for executing an auto starting in VRAK was to simply disengage VRAK during the beginning of the autorotation. Pilot B suggested that a future version could "kick off VRAK when NR splits." The research team agrees that additional design work is warranted.

The major takeaway, however, was well said by Pilot C: "Yes, you can do an auto with VRAK."

TERF Flight. Pilots were asked to fly the terrain following (TERF) course at NAS Patuxent River as simulated in CASTLE to evaluate VRC in a non-precision/aggressive piloting task. The Pax River TERF course has moderate elevation changes (due to trees and coastline) and moderately tight to very tight turns. Engineers are able to specifically tune controls to perform well for precision tasks since those tasks typically have success criteria (or tolerances) that are easily translated into quantitative metrics for evaluation. Non-precision flight, in contrast, does not always have those clear criteria and, therefore, are a more challenging design task. Accordingly, the research team was especially interested in the results from this evaluation of VRC. The pilots were instructed to select and transition flight augmentations at their own discretion.

The results of the TERF flight were extremely informative. There were two major results from the testing. One related to the collective backdrive behavior and the other related to the interaction between VRAK and PFCS.

Pilots C and B had opposite opinions of the collective backdrive. Pilot C found VRAK's backdrive desirable while Pilot B found it objectionable. Executing a turn, Pilot C felt the collective rise in their hand to maintain altitude and they liked it. They said "It feels like the collective is doing what it is supposed to do... when I roll to turn, if I didn't feel the collective coming up, I would be compelled to pull it up myself." This comment by Pilot C confirms the research team's hypothesis that backdriving the position of the collective to match the true collective blade pitch does indeed increase confidence in the mode for some pilots. In contrast, flying a similar turn, Pilot B felt the collective rise in their hand and they disliked it. They said, "If I feel the collective move, my inclination is to stop it." This behavior, whether considered a desirable or objectionable feature, is inherent to the VRAK implementation of VRC, not in VRC itself. These contrasting opinions on basic inceptor behavior are detailed in the Implications for Tactile Envelope Cueing subsection of Results.

Pilot A had a third opinion. They found it to be "not detrimental," but "the biggest shift in control strategy" between all of the test events from that of BL. They found that the backdrive felt slow when performing rolls and other aggressive maneuvers on the cyclic. The backdrive performance can be improved by tuning its parameters and developing the trim map. The trim map is currently parameterized on ground speed and altitude and feedback is used to account for unmodeled dependencies. Expanding the trim map to include dependence on other states, such as attitude, could improve the backdrive performance reducing the dependence on feedback.

During their flights with VRAK+PFCS, Pilot A noticed that the aircraft had a tendency to "run away" with airspeed. Without hawking the flight display, flying in a manner that in BL would hold an airspeed of around 80 kts, Pilot B found themselves easily reaching 140 kts. The research team speculates that this behavior may be rooted in undesired interactions between VRC and PFCS's rate command/attitude hold (RCAH) cyclic. If VRC is, indeed shown to pair poorly with RCAH, then a PFCS specific vertical response augmentation may be desirable. This subject should be pursued in future work.

Pilot comments indicate that, in its current state, VRAK's realization of VRC is not beneficial for TERF. It does not prevent TERF, but there are no clear benefits and further tuning of the inceptor backdrive and improvements to the trim map appear to be warranted.

Pilot B's comments and the mode's tendency to accelerate undesirably indicate that VRC may not be the preferred control method for all phases of flight (i.e. better as a task selectable pilot mode). These are limited results and are not sufficient to determine whether VRC is robust enough to be a full time control mode. This question should be the subject of future research.

Implementation-Practicality

The implementation-practicality implications of VRAK and SFD were assessed in an independent engineering evaluation conducted in the CH-53K™ flight simulator at NAS Patuxent River. This assessment demonstrated that VRAK's SFD control method addresses each of the three challenges to implementation-practicality. Additional results pertaining to the implications of SFD are drawn based on pilot comments.

Accommodating Mode Transition Responses. MTRs are most likely to be undesirable or even objectionable in circumstances such as landing, low-altitude flight, and TERF flight, where an undesired vertical response could lead to a mishap. SFD is designed to handle MTRs in such situations by preventing them.

Engaging VRAK from BL, the inceptor begins in a position and rotor state already consistent in terms of S2H-effect. VRAK initializes FPI and the trim map based on that initial condition leading to no MTR.

Disengaging VRAK into BL is the more challenging case from a control design perspective. In steady-state, SFD is designed to match the position of the collective between VRAK and BL so that when the pilot disengages VRAK into BL, the control signal to the rotor head would not change. With no change in rotor, there would be no MTR.

SFD's mitigation of MTRs was demonstrated during an independent engineering flight in the CH-53K™ flight simulator. Three cases were considered. In order of increasing difficulty, they were:

1. Disengaging VRAK into BL with C-AIS in detent
2. Disengaging VRAK into BL with C-AIS outside of detent at rest
3. Disengaging VRAK into BL with C-AIS outside of detent in motion

Based on SFD's design, the research team expected SFD to eliminate the MTRs in the first two cases and to reduce the MTR to a mode transition transient (MTT) in the third case.

SFD successfully eliminated MTRs in the first two cases and successfully reduced the MTR to a maximum of 0.2 g MTT in vertical acceleration in the third case. This transient would most likely not be objectionable to a pilot as it would only represent a change in a non-zero vertical speed, not a transient from zero. If it were, the fact that SFD reduced it from an MTR to a MTT means it would now be possible to address the issue with a simple approach such as a transient free switch.

The vertical rate and collective motion responses from the second case are illustrated in Figure 8. For Figure 8.a, which shows the vertical rate response, the dashed black line shows the commanded vertical rate, the red line shows the reference vertical rate, and the blue line shows the response vertical rate. For Figure 8.b, which shows the collective motion response, the solid black line shows the real world motion of the collective, the blue line shows the command to the collective, and the dashed cyan and magenta lines show the collective's detent. When VRAK disengages around 50 seconds, the collective is held so that it does not move. With the S2H-effect of inceptor position matched between VRAK and BL, the vertical rate of the aircraft does not change due to the transition. The lack of an MTR demonstrates SFD's practicality.

Preserving Full Control Power. In the interest of safety, vertical augmentation needs to preserve pilots full control power so that they are always able to pull full power or autorotate should the need arise. VRAK is designed to preserve control power through the force cutoffs noted in the Command Map subsection of VRAK Architecture. These force cutoffs are a practical approach when paired with SFD owing to SFD's commanding vertical rate based on force. SFD changes the way force moves the collective, but if the pilot pushes or pulls more the threshold, they will move the collective through one of its cutoffs. Past these cutoffs, VRAK is disengaged and the collective behaves exactly as it would in BL. The pilot is always, therefore, able to pull full collective to command full power and put the collective to the floor to autorotate.

Figure 9 shows the motion of the collective through its full throw with a reasonable amount of force. For Figure 9.a, which shows the collective motion response, the solid black line shows the real world motion of the collective, the blue line shows the command to the collective, the dashed cyan and magenta lines show the collective's detent, and the red line shows the state of VRAK—VRC or S2H. In this test, the inceptor was moved through its full range of motion. VRAK

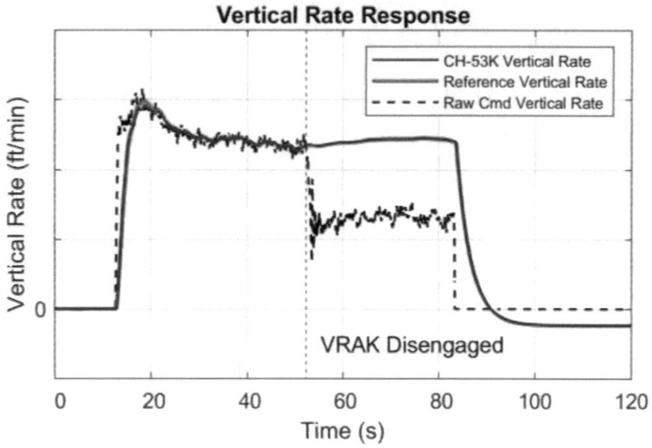

Figure 8.a Vertical rate response to disengaging VRAK and releasing to detent

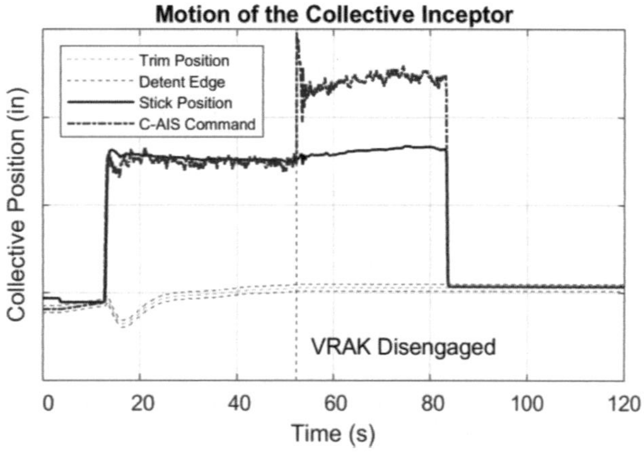

Figure 8.b Collective motion response to disengaging VRAK and releasing to detent

Figure 8. Vertical rate and collective motion responses to MTR case 2 and to a release to detent starting from 25 kts forward air speed at 100 ft altitude

was engaged initially and, when the system detected that the pilot was pulling out of VRAK's limits, the mode automatically suspended VRC, switching to S2H. The pilot never manually downshifted from VRAK to BL. When the pilot brought the C-AIS back towards detent, VRAK automatically returned to VRC. The minor transients in force (and commanded C-AIS position) are due to pulling the inceptor against the stops at the end of its throw. These results demonstrate that VRAK does not prevent pilots from commanding full control power.

Managing the Collective's Non-Unique Trim. When engaging and disengaging vertical augmentation, the control system must manage the collective detent to ensure predictability and desirable behavior.

VRAK is designed to account for the collective's non-unique trim through SFD and its trim map. While engaged, VRAK does not use a detent in the same way as the BL system. The inceptor position is driven independently of the detent location using the AIS. While engaged, for convenience, VRAK places the detent at the inceptor position corresponding to the trim vertical rate, as calculated by the trim map. Upon mode disengage into BL, the collective's detent is already in the position corresponding to VRAK's trim vertical rate. With VRAK engaged, the position of the detent is updated as the aircraft changes flight condition and as feedback corrects any low-frequency errors in the collective position.

VRAK's detent management is demonstrated in Figure 8, alongside MTR rejection. At the end of the test sequence, the pilot has disengaged VRAK with the inceptor out of detent and they release the stick. The stick falls to the detent that was frozen at the time VRAK disengaged. The vertical rate converges to a slow descent. The aircraft did not come to a hover, as would be most desirable with trim set to zero vertical rate in VRAK. A small residual, however, is not unexpected. The aircraft climbed substantially between the time that VRAK was disengaged and the time that the collective was released. The changing flight condition, therefore, led to the slight descent with the stick in detent. The behavior is most likely not objectionable.

Discussion of Implementation-Practicality. These results demonstrate that the SFD control method strategy of maintaining a steady-state one-to-one correspondence between inceptor position and collective blade pitch angle, along with several other enabled features, addresses the three challenges to implementation-practicality. The fact that the simulator-integrated SFD control method, not just its design, fulfills these three claims establishes SFD as a real, practical solution the implementation challenges. Demonstrating a practical approach to these three challenges increases the likelihood of a successful and beneficial real implementation of vertical augmentation on a real development or production rotorcraft.

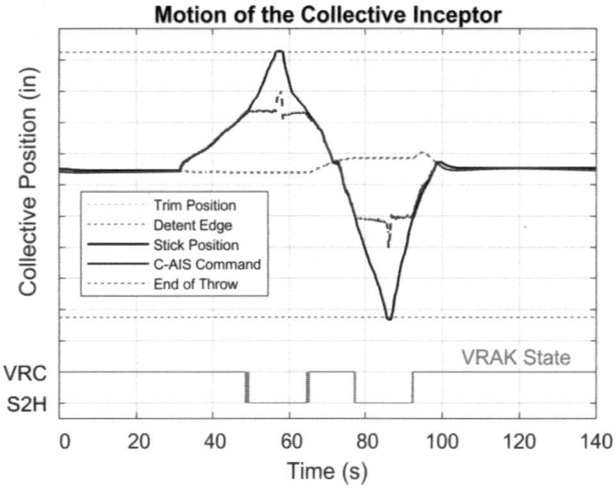

Figure 9.a Motion of the C-AIS and VRAK engagement state

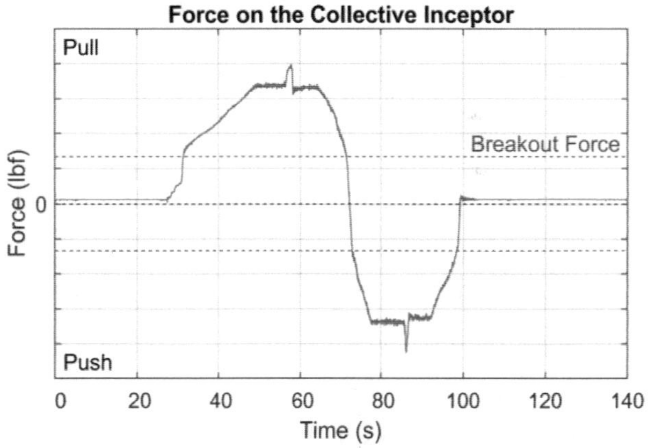

Figure 9.b Force applied to the C-AIS

Figure 9. Moving the C-AIS through its full throw beginning with VRAK engaged and without downshifting from VRAK

Implications for Tactile Envelope Cueing

A system with tactile envelope cueing is one that provides pilots information on the current state of the aircraft via proprioceptive and haptic feedback, thereby offloading that information from the visual and auditory channels, which are incredibly overtaxed in aviation. VRAK provides pilots with envelope cueing through its backdrive of the collective. For example, starting from steady-level flight, pilots are able to feel the aircraft decrease collective to maintain altitude when they pitch their nose back.

Implications of this tactile envelope cueing were highlighted during the TERF Flight section of the pilot evaluation. During TEFF, Pilots B and C commented opposite opinions on the collective backdrive. Their comments are reproduced here. After flying a turn, Pilot C commented "It feels like the collective is doing what it is supposed to do… when I roll to turn, if I didn't feel the collective coming up, I would be compelled to pull it up myself." Flying a similar turn, Pilot B commented "If I feel the collective move, my inclination is to stop it." During the turns, as the pilots rolled over and the thrust required to maintain zero vertical rate increased, VRAK drove the collective to rise in the pilots hands. Pilot C interpreted the pressure on and motion of their hand as a cue indicating the control system was operating correctly. Pilot B interpreted the same sensations as a cue indicating the control system was operating incorrectly. Pilot C preferred having greater insight into what the system was doing and Pilot B wanted the collective to stay fixed if the control system was holding vertical rate, similarly, fixed.

Pilot C's comments illustrate one implication and benefit of envelope cueing: increasing pilot confidence in augmentation. Pilot C was much more confident flying VRAK knowing what he could look for to tell if the system was not behaving as expected. The flipside to this benefit should be noted as well: envelope cueing can also decrease pilot confidence in higher augmentation.

Pilot B's comment presents an important topic for future work. Pilot B said that if they are holding a condition and the collective moves, they want to command a return to their held condition by stopping the motion of the inceptor. For example, if the aircraft starts to incorrectly descend, they want the collective to fall so that the stick will cue them to pull up to arrest the incorrect descent. This behavior is consistent with a link between inceptor position and *real world* vertical rate. If the aircraft starts descending, using this scheme, the stick will fall to indicate that descent. Focused on something else, the pilot might naturally resist the deflection and the spring action of their arm may naturally return the stick to detent and, in so doing, return the aircraft to the trim vertical rate. This scheme is an example of envelope cueing since the motion of the collective is linked to a *measurement* of the aircraft's flight condition and not to a *command*. In this scheme, the command signal for vertical rate would likely need to be linked to inceptor force, similar to VRAK. Such a scheme would be an extension of other works in which inceptor position is used to command vertical rate (Refs. 4, 5). This scheme would lack the practicality advantages of VRAK's link between inceptor position and collective blade pitch, but those losses might be offset by the benefits of envelope cueing. The research team recommends researching this design candidate in future work.

Beyond the test pilot evaluations of VRC, the research team also performed an informal human factors evaluation of VRAK with Pilot D. Pilot D is a human factors SME with a Ph.D. in Industrial Engineering; Aviation Human Factors and currently teaches Human Factors at the USN Test Pilot School at NAS Patuxent River. During the human factors evaluation of VRAK, Pilot D identified an unintended implication of the SFD technology: simple and effective envelope cueing.

In Pilot D's professional opinion, SFD is a "powerful" and innovative implementation of envelope cueing technology. He elaborated:

> "I've been talking about envelope cueing for the last ten or fifteen years and this is the first time where I've seen it implemented in a way that was well implemented and provided a lot of information with a relatively simple system... you didn't have buzzers and beepers and all this stuff, you just had the collective moving and, without demanding changes to vertical climb, I was able to tell what my power was doing once I set it because of how the collective backdrove. It was really nice to see."

SFD offers a continuous low-frequency envelope cue on collective blade pitch. This cue enables pilot awareness of controls being applied independently by the control laws. This awareness is desirable in high gain situations where a pilot's actions might depend on, for example, the amount of remaining power available.

The cue provided by SFD is clear and useful to pilots. Similar architectures, built around active inceptors, can offer similar capabilities for other variables of interest. SFD's continuous envelope cueing capability makes the technology a promising prototype for next generation rotorcraft envelope cueing in the vertical axis.

CONCLUSIONS

This paper described the development of the Vertical Response Augmentation for CH-53K™ prototype control augmentation which effects a vertical rate command response type and addresses several key challenges to the implementation-practicality of rotorcraft vertical augmentation. Through a formal handling quality evaluation, VRAK's realization of the VRC response type was demonstrated to have workload and handling quality benefits in a wide range of tasks. Through engineer evaluation, VRAK was demonstrated to address three challenges to implementation-practicality. Further implications for tactile envelope cueing were also identified. The following conclusions are reached:

1. Significant reductions in workload and improvements in handling qualities as a result of VRC were observed through improved HQRs and positive pilot comments in the Vertical Maneuver MTE and Aerial Refueling.
2. Some reduction in workload and improvements in handling qualities as a result of VRC were observed through pilot comments in the Precision Hover MTE.
3. Small to negligible benefits to workload and handling qualities as a result of VRC were observed through pilot comments in the Pattern. This was expected.
4. Pilot comments indicate VRAK's realization of VRC is, in its current state, not beneficial for TERF. It does not prevent TERF, but there are no clear benefits. Further tuning of the inceptor backdrive and improvements to the trim map appear to be warranted.
5. VRC did increase the workload associated with Nr control in autorotations. It, however, did not impact the ability to safely execute the autorotation.
6. VRAK's SFD control method was demonstrated to be a practical solution to the implementation challenge posed by MTRs on the collective.
7. VRAK's force cutoffs, enabled by SFD, were demonstrated to be a practical solution to the implementation challenge of preserving full collective control power.
8. VRAK's trim map was demonstrated to be a practical solution to the implementation challenge of managing the collective detent through mode transition.
9. SFD has a continuous envelope cueing capability which makes the technology a promising prototype for next generation rotorcraft vertical envelope cueing.

Author contact: Ian Willebeek-LeMair
ian.j.willebeek-lemair.civ@us.navy.mil
Matthew Rhinehart matthew.h.rhinehart2.civ@us.navy.mil

ACKNOWLEDGMENTS

This project was funded by the Naval Innovative Science & Engineering (NISE) program. The authors wish to thank Eric O'Neill, Alexandria Songy, Dan Shafer, and Dave Engel for extensive support throughout the research, evaluation, and documentation efforts and to thank the CH-53K™ test pilots and Pilot D for evaluating the prototype control law.

REFERENCES

1. Anon., "Handling Qualities Requirements for Military Rotorcraft," U.S. Army Aviation and Missile Command ADS-33E-PRF, March, 2000.

2. Spires, J. M., and Horn, J. F., "Multi-Input Multi-Output Model-Following Control Design Methods for Rotorcraft," American Helicopter Society 71st Annual Forum Proceedings, Virginia Beach, VA, May 2015.

3. Harding, J. W., Moody, S. J., Jeram, G. J., Mansur, M. H., and Tischler, M. B., "Development of Modern Control Laws for the AH-64D in Hover/Low Speed Flight," American Helicopter Society 62nd Annual Forum Proceedings, Phoenix, AZ, May 2006.

4. Berger, T., Blanken, C. L., Lusardi, J. A., Tischler, M. B., and Horn, J. F., "Coaxial-Compound Helicopter Flight Control Design and High-Speed Handling Qualities Assessment," *Journal of the American Helicopter Society*, Vol. 67, (3), Jul. 2022, pp. 98-113(16). DOI: 10.4050/JAHS.67.032008.

5. Berger, T., Blanken, C. L., Lusardi, J. A., Tischler, M. B., and Horn, J. F., "Tiltrotor Flight Control Design and High-Speed Handling Qualities Assessment," *Journal of the American Helicopter Society*, Vol. 67, (3), Jul. 2022, pp. 114-128(15). DOI: 10.4050/JAHS.67.032009.

6. Cooper, George E., Harper, Robert P. Jr., "The Use of Pilot Rating in Evaluation of Aircraft Handling Qualities", Advisory Group for Aerospace Research and Development, Report 567, April 1967.

Predicting Pilot Performance, Workload and Handling Qualities using SCOPE

Edward Bachelder
Senior Research Engineer
San Jose State University
Moffett Field, CA, USA

Bimal Aponso
Chief Scientific & Technical
Advisor for Dynamic Systems
Federal Aviation Administration
Seattle, WA, USA

Tom Berger
Flight Controls Group Lead
U.S. Army CCDC-AvMC
Moffett Field, CA, USA

ABSTRACT

The Cooper-Harper scale provides a structured approach to assessing handling qualities. It employs a decision tree structure usings various descriptors whose meaning and practical application could be misunderstood or overlooked. In this scale the term 'pilot compensation' is a key discriminator, but its definition, as offered by Cooper and Harper, is weakly defined and difficult to quantify. Previous research on estimating Handling Qualities Rating (HQR) from flight data typically focused on correlating pilot-vehicle performance measures with the HQR ratings. The approach presented in this work estimates workload and HQRs by establishing baseline workload levels for a given task using reference vehicle command dynamics. These baseline levels are generated by the Spare Capacity OPerator Estimator (SCOPE), which estimates workload with a model of pilot neuromuscular feedback, a cost function that captures pilot behavior, and the stick-to-disturbance sensitivity transfer function. Using the baseline workload levels and the original definitions found in the HQR scale, quantitative values from SCOPE are mapped to the HQR descriptors 'compensation' and 'controllability'. The proposed approach allows designers to predict task performance and provide diagnostic and prognostic workload and HQR capability. A piloted simulation study using two Future Vertical Lift-relevant rotorcraft configurations (a lift offset coaxial helicopter with a pusher propeller and a tiltrotor) was conducted. SCOPE's predictive performance and HQR estimates for the study's pitch attitude capture and hold task compared favorably with the actual pilot performance and ratings. SCOPE did not employ pilot data for calibrating its estimates.

NOTATION

B	=	Bedford workload rating
d	=	Forcing function signal
CL	=	Closed loop
e	=	Error signal
f	=	Force
FS	=	Feel system
G_{xy}	=	Transfer function, x output y input
G_{NM}^{OL}	=	Pilot OL NM transfer function
HQR1	=	Handling Qualities Rating of 1
K_P	=	Proprioceptive pilot gain (steady-state)
K_V	=	Vehicle gain (steady-state)
K_D	=	Display gain (steady-state)
NM	=	Neuromuscular
OL	=	Open loop
PF	=	Proprioceptive feedback
PSD	=	Power spectral density
s	=	Complex variable, Laplace transform variable
SCOPE	=	Spare Capacity OPerations Estimator
S_{xx}	=	PSD of signal x
t	=	Time
δ	=	Stick signal
μ	=	Workload stimulus
μ^{γ}	=	Workload metric
σ_x^2	=	Variance (power) of signal x
ζ	=	Damping ratio
τ	=	Pilot time delay (sec)
γ	=	Perceptual exponent
ω	=	Angular frequency, rad/s
ω_c	=	Crossover frequency, rad/s
ω_{Ld}	=	Pilot proprioceptive lead frequency, rad/s
ω_{Lg}	=	Pilot proprioceptive lag frequency, rad/s

INTRODUCTION

The Cooper-Harper scale provides a structured approach to assessing handling qualities. It employs a decision tree structure usings various descriptors whose meaning and practical application could be misunderstood or overlooked. In this scale the term 'pilot compensation' is a key discriminator, but its definition, as offered by Cooper and Harper, is weakly defined and difficult to quantify. Previous research on estimating Handling Qualities Rating (HQR) from flight data typically focused on correlating pilot-vehicle performance measures with the HQR ratings. The approach presented in this work estimates workload and HQRs by establishing baseline workload levels for a given task using reference vehicle command dynamics. These baseline levels are generated by the Spare Capacity OPerator Estimator (SCOPE), which estimates workload with a model of pilot

Presented at the VFS Handling Qualities Technical Meeting, Huntsville, Alabama, USA Feb. 7-8, 2023.

neuromuscular feedback, a cost function that captures pilot behavior, and the stick-to-disturbance sensitivity transfer function. Using the baseline workload levels and the original definitions found in the HQR scale, quantitative values from SCOPE are mapped to the HQR descriptors 'compensation' and 'controllability'. The proposed approach allows designers to predict task performance and provide diagnostic and prognostic workload and HQR capability. A piloted simulation study using two Future Vertical Lift-relevant rotorcraft configurations (a lift offset coaxial helicopter with a pusher propeller and a tiltrotor) was conducted. SCOPE's predictive performance and HQR estimates for the study's pitch attitude capture and hold task compared favorably with the actual pilot performance and ratings. SCOPE did not employ pilot data for calibrating its estimates.

OVERVIEW OF HANDLING QUALITIES RATING AND ITS FUNCTIONAL COMPONENTS

The Cooper-Harper Handling Qualities Rating (HQR) scale is the accepted standard scale for assessing handling qualities. The capability to analytically forecast pilot performance, workload, and handling qualities rating (HQR) with high fidelity does not currently exist. ADS-33 (Ref. 1) provides predictive metrics for Handling Qualities Level 1 – 3 (representing HQR groupings), and the pilot is not explicitly accounted for. Reference 2 examined reasons for variability in qualitative pilot HQRs and concluded much of the inherent variability is directly attributable to the metathetic nature of handling qualities, in conjunction with the application of an ordinal rating scale. Reference 3 reviewed factors more related to human nature and experimental setup that can impinge on the pilot and adversely influence the subjective rating. These factors include: 1) Proclivities and vested interests (aircraft appearance, financial impact, potential for gain or loss of status); 2) Improper mindset (the correct approach is that it is the aircraft being evaluated, not the pilot); and 3) Unrealistic performance standards relative to the actual task - the desired and adequate performance limits should reflect operational needs. For example, desired performance for a landing task would dictate minimal use of braking, whereas adequate performance means that touchdown should occur on the runway and with enough runway left to stop. One of the key points in Ref. 3 is that workload should always be emphasized over performance when assigning the ratings - performance is included primarily to standardize the task aggressiveness.

Reference 4 provides a summary of previous research examining measures that correlate with pilot workload and Cooper Harper Handling Qualities Ratings (HQRs). A weighted adaptive attack rate metric is proposed in Ref. 4 that correlates well with handling qualities ratings for a range of mission task elements (MTEs). Rather than investigating correlation between HQRs and task metrics, the approach presented herein estimates HQRs by first predicting pilot performance and the associated workload and phase margin for two reference conditions based on desired and adequate performance. These are then employed to populate the HQR rating scale using the textual descriptors found in the HQR scale. Once the HQR scale is complete, the pilot's performance, workload, and phase margin are estimated. A key objective of this work is to help pilots and researchers understand the physical meaning of compensation, its role in HQR, and the distinction between workload and compensation. The proposed approach will enable designers to predict task performance and provide workload and HQR diagnostic/prognostic capability.

In Reference 5 Cooper and Harper state *Pilot evaluation still remains the only method of assessing the interactions between pilot-vehicle performance and total workload in determining suitability of an airplane for the mission. It provides a basic measure of quality and serves as a standard with which pilot-airplane system theory may be developed, performance measurements may be correlated, and significant airplane design parameters and characteristics may be determined and correlated.*

The Cooper Harper HQR scale (Figure 1) employs a decision tree that is based on descriptors that include *task*, *workload*, *compensation*, *performance*, and *control*.

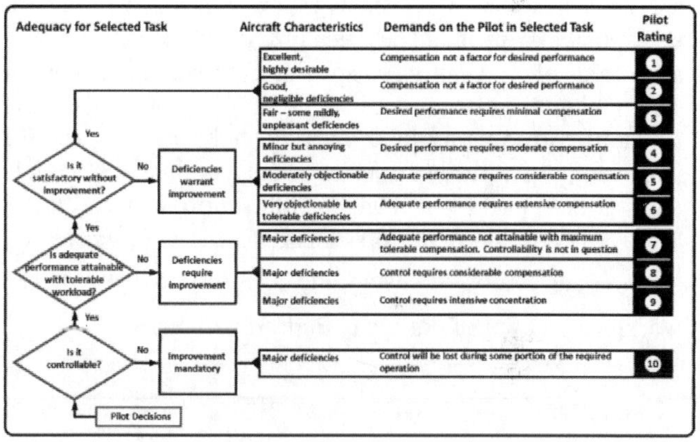

Figure 1. Cooper Harper handling qualities rating scale (Ref. 5).

Task

A task is defined by maneuver, target states, forcing function (external disturbance), and performance criteria.

Workload

Figure 3 shows a block diagram of the compensatory tracking task, where Y_P, G_{FS}, and G_V are the pilot, feel system, and vehicle elements, respectively. In Reference 6 Bachelder observed that the pilot workload associated with compensatory control was linearly related to error rate and stick rate, from which the Spare Capacity OPerations Estimator (SCOPE) was developed as an estimator of workload. SCOPE computes an estimated Bedford workload rating B_{est} from the product of the standard deviations of error rate and control rate raised to an exponent γ, shown in Eq. 1.

$$B_{est} = B_o + \frac{dB}{d\mu^\gamma}(\sigma_{\dot{e}}\sigma_{\dot{\delta}})^\gamma = B_o + \frac{dB}{d\mu^\gamma}\mu^\gamma \qquad 1$$

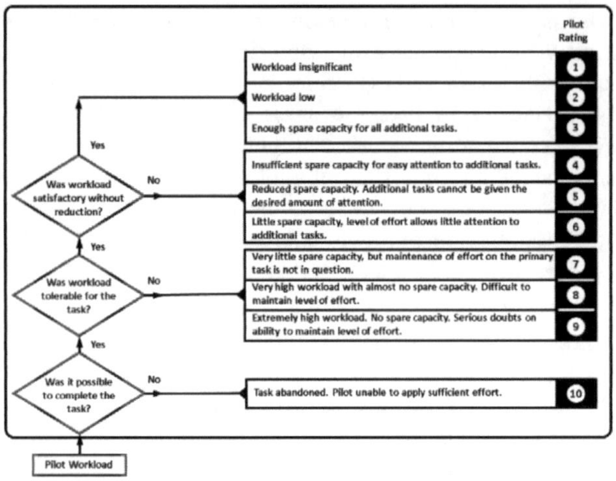

Figure 2. Bedford workload rating scale (Ref. Error! Bookmark not defined.**).**

The product in parenthesis $(\sigma_{\dot{e}}\sigma_{\dot{\delta}})$ is referred to as the workload stimulus μ, and μ^{γ} is denoted as the SCOPE workload metric. B_o is the B_{est} intercept when $\mu^{\gamma} = 0$, and $\frac{dB}{d\mu^{\gamma}}$ is the Bedford-to-workload metric gradient (B_o and $\frac{dB}{d\mu^{\gamma}}$ are constants that are task-specific). γ is denoted as the perceptual exponent and has been observed to have a value of approximately 0.07 for compensatory tracking. Figure 4 shows high correlation between SCOPE's workload metric μ^{γ} and Bedford rating for a tracking study (Ref. 6) where a wide range of vehicle dynamics, display gains, and vehicle gains were used. In Figure 5 the workload metric is transformed to the Bedford estimate using Eq. 1, where $\boldsymbol{B_o}$ and $\boldsymbol{\frac{dB}{d\mu^{\gamma}}}$ are computed using linear regression between the workload metric and the actual Bedford ratings.

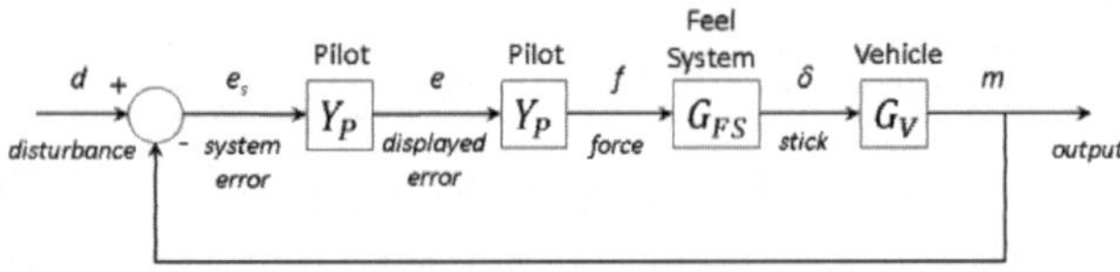

Figure 3. Compensatory tracking task signals and elements.

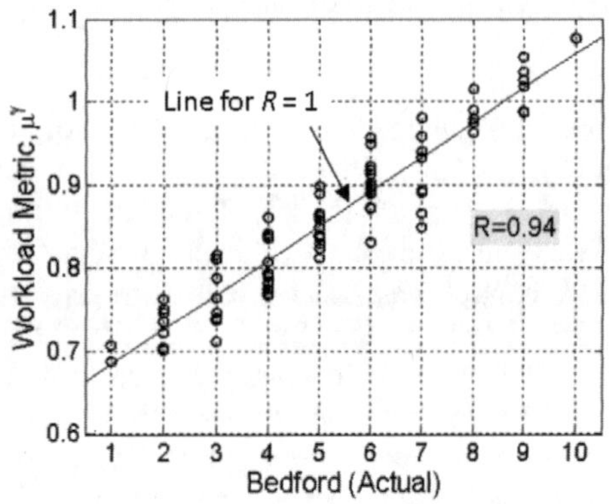

Figure 4. SCOPE workload metric versus Bedford rating (N = 81, correlation coefficient = 0.94, Ref. 6).

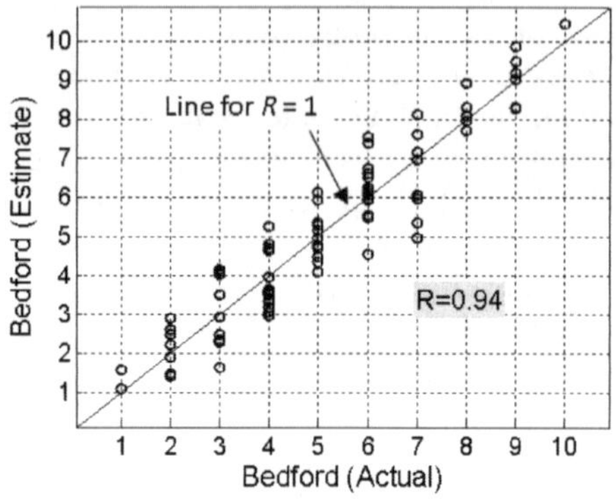

Figure 5. SCOPE Bedford estimate versus actual Bedford ratings (N = 81, Ref. 6).

Until now the challenge has been predicting the task-specific B_o and $\frac{dB}{d\mu^{\gamma}}$ to enable Bedford workload estimation using measured workload stimuli, thus eliminating the requirement for actual workload ratings for calibration. Reference 16 provides such a data-free calibration approach.

In Reference 7 Bachelder extended fundamental laws of sensory perception (single input) to multi-input sensing, showing that multiple sensory inputs add logarithmically (equivalently, they multiply via the power law). It was proposed rather than being just a sensational response, a pilot's perception of workload is a skill that is acquired and used to modulate system stability (i.e., phase margin) during a tracking task.

The relationship in Eq. 1 assumes the inceptor force is roughly proportional to stick deflection, i.e., the stick's spring-centering force is much larger than the force effects due to inertia and damping. When this assumption is not valid, the workload stimulus takes the general form

$$\mu = \sigma_{\dot{e}}\left[\sigma_{\dot{\delta}}\sigma_{\dot{f}}\right]^{\frac{1}{2}} \qquad 2$$

When two or more perceptual stimuli are generated by the same source, the exponent applied to each of the inputs is the reciprocal of the number of stimuli. The same limb produces and senses both f and δ, hence each stimulus is assigned the exponent of 1/2 (Ref. 8). Cooper and Harper (Ref. 5) defined workload as including both mental and physical effort. SCOPE captures mental effort through the observed error rate and physical effort through stick deflection rate and stick force rate.

A Taylor series expansion of the workload metric μ^{γ} evaluated at $\gamma = 0$ yields Eq. **3**

$$\mu^{\gamma} \approx 1 + \gamma \log \mu \quad \text{for } \gamma \ll 1 \qquad 3$$

Thus, for small values of γ the power function resembles the logarithmic function. Figure 6 compares the response of the power function with the logarithmic approximation in Eq. 3 over a typical range of $\boldsymbol{\mu}$ using $\gamma = 0.07$, and the approximation is seen to have a 1% error at $\boldsymbol{\mu} = 8$.

Figure 7 illustrates the logarithmic nature of the Bedford rating relative to the workload stimulus μ. When μ is low, the pilot can perceive each unit increase in rating with high sensitivity to the stimulus (little change in μ is required to sense a change in workload). When the stimulus values are high, sensitivity to the stimulus is substantially reduced, requiring a large change in μ to perceive a unit change in the Bedford scale.

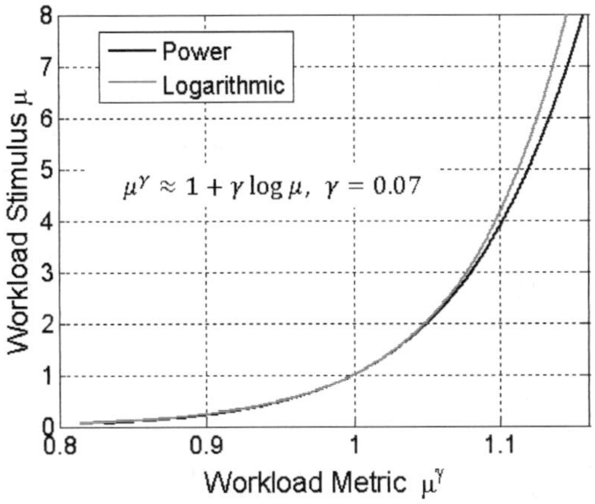

Figure 6. Workload stimulus μ versus workload metric μ^{γ}.

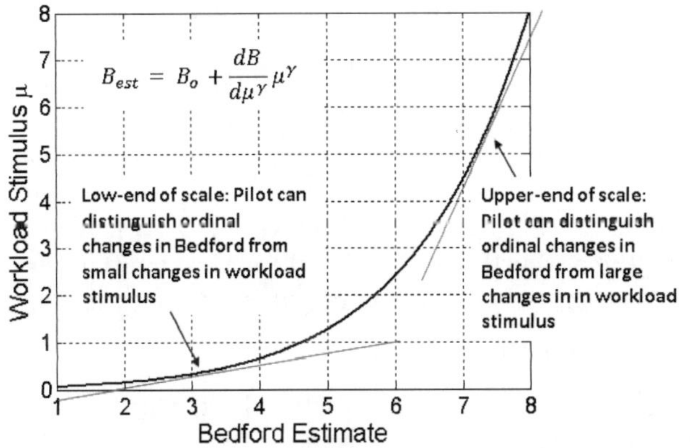

Figure 7. Workload stimulus μ versus Bedford estimate.

Compensation

Cooper and Harper (Ref. 5) defined compensation as the *measure of additional pilot effort and attention required to maintain a given level of performance in the face of deficient vehicle characteristics. The total workload is comprised of the workload due to compensation for aircraft deficiencies plus the workload due to the task.* Compensation is thus relative to some baseline reference that depends only on the task – this is a critical detail. As an example, assume a tracking task where a commanded pitch symbol is driven by a series of step changes. For fixed inceptor and display dynamics, the baseline reference condition would employ aircraft (command) dynamics such that any other command dynamics would produce higher workload. In References 7 and 9 Bachelder demonstrated that proportional command dynamics yield the lowest workload and tracking error compared to rate and acceleration command systems. In general, however, pilots are more accustomed to aircraft with rate command dynamics, where the stick is returned toward the neutral position (zero force when the stick is trimmed) as the aircraft approaches the desired attitude. This contrasts with the control strategy employed with proportional command dynamics, where stick motion is ceased as the desired attitude is approached. From 1985-1990 the U.S. Air Force (USAF) conducted a comprehensive study on Minimum Flying Qualities (Refs. 2, 10, and 11) that showed rate-command dynamics produces Level 1 flying qualities (HQR1 – 3). Based on this finding, and control style familiarity, the command dynamics selected for generating the baseline compensation associated with HQR1 (HQ rating of 1) was rate-command.

Since compensation is relative to a baseline workload, absolute workload (such as Bedford rating) is not required. Instead, only the workload metric μ^{γ} which is comprised of the observable measures error rate and stick rate (Eq. 1) is needed. This is shown in Figure 8 where the baseline Bedford workload associated with an HQR1 is denoted as B_{HQR1}, which is linearly related to the SCOPE workload metric ${\mu^{\gamma}}_{\mathrm{HQR1}}$ (Eq. 1). As discussed later, the second baseline used to calibrate compensation is the workload associated with HQR6 (B_{HQR6}). The implicit assumption is that the compensation descriptors for HQRs 1 through 6 represent compensation that increases linearly with HQR. Figure 8 shows that interpolating between the two baseline workloads to compute $B_{\mathrm{HQR}N}$ (handling qualities rating N where $1<N<6$) is equivalent to interpolating between the two baseline workload metrics ${\mu^{\gamma}}_{\mathrm{HQR1}}$ and ${\mu^{\gamma}}_{\mathrm{HQR6}}$ to compute ${\mu^{\gamma}}_{\mathrm{HQR}N}$. Note that μ^{γ} is used as a proxy for compensation instead of the workload stimulus μ because μ^{γ} is linearly related to workload (hence compensation), as seen in Figure 3b.

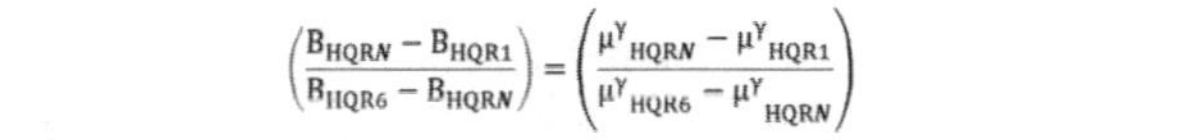

Figure 8. Similitude between workload associated with compensation and SCOPE workload metric.

Performance

The HQR scale categorizes performance as 'desirable' and 'adequate', which is defined by HQ task performance requirements that are based on aircraft states observable by the pilot.

Figure 9 shows the closed loop pilot-vehicle system and associated four attributes (denoted by red arrows) that are used as inputs for HQR evaluation. It would be more accurate to place the Display element between the Task and Pilot elements, but for ease of exposition it is combined with the Vehicle and Feel System elements.

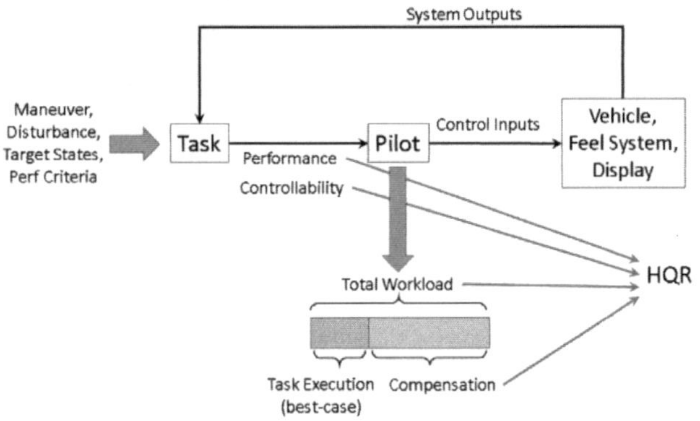

Figure 9. Closed loop system attributes (denoted by red arrows) used as inputs for HQR evaluation.

MODIFIED STRUCTURAL MODEL

Performance and workload are estimated using a modification of Hess' Structural Model (Ref. 12) shown in Figure 10, denoted as the Modified Structural Model (MSM). The open-loop NM system G_{NM}^{OL} is represented as a second order system with a pure time delay,

$$G_{NM}^{OL} = \frac{{\omega_{NM}^{OL}}^2 e^{-\tau_{NM}s}}{(s^2 + 2\zeta_{NM}^{OL}\omega_{NM}^{OL} + {\omega_{NM}^{OL}}^2)} \quad 4$$

G_{NM}^{OL} has typically been modelled in manual control as having the mode ω_{NM}^{OL} residing at 10 rad/s with a damping ratio ζ_{NM}^{OL} of 0.7. However, in Ref. 13 Bachelder observed ω_{NM}^{OL} to vary significantly depending on the command type, ranging from approximately 1.5 rad/s for proportional command to 12 rad/s for acceleration command.

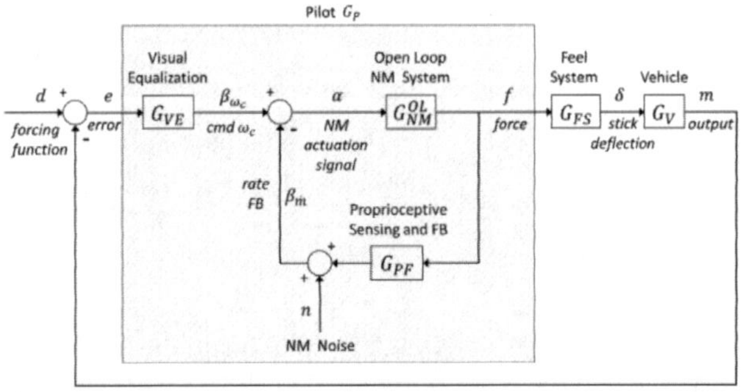

Figure 10. Modified Structural Model (MSM) for force feedback control.

Departing from the Structural Model (SM) which assumes infinite lead or lag depending on the vehicle dynamics G_V, proprioceptive sensing and feedback element G_{PF} is modeled as a generic first-order lead-lag with a proportional gain K_{PF}, where both proprioceptive lead and lag are limited to the closed loop NM frequency ω_{NM}^{CL} (Ref. 13),

$$G_{PF} = K_{PF}\frac{\omega_{Lg}(s+\omega_{Ld})}{\omega_{Ld}(s+\omega_{Lg})} \quad 5$$

The SM assumes stick deflection is used for the NM feedback signal, and that the feel system (G_{FS}) is included in the NM control loop. This provides the degree of freedom to shift ω_{NM}^{CL} below the open loop NM frequency (ω_{NM}^{OL}). However, in Ref. 9 Bachelder presented evidence that force (not stick deflection) is predominantly employed for control feedback when the inceptor is spring centering, i.e., the feel system lies outside the NM loop. Without the feel system in the loop, NM feedback will produce a ω_{NM}^{CL} that is greater than or at least equal to ω_{NM}^{OL}, i.e., when ω_{NM}^{CL} has been observed to be less than 10 rad/s ω_{NM}^{OL} must likewise be less than 10 rad/s. This is consistent with the experimental results of Ref. 13.

Also shown in Figure 10 is NM noise arising from proprioceptive sensing and feedback signal generation (noise assumed to be zero for the current work). This is added to the linear output of G_{PF} producing the NM feedback signal $\beta_{\dot{m}}$ which is effectively inner-loop rate feedback. This rate feedback serves to condition the output m such that its rate $\dot{m}$ is proportional to the error signal e (i.e., the system conforms to the Crossover Model). The visual equalization element G_{VE} is represented as a gain K_e and pure time delay τ_e. This delay is associated with visual and central nervous system processing (Ref. 14)

$$G_{VE} = K_e e^{-\tau_e s} \quad 6$$

Hess' SM assumes τ_e to be fixed at 0.2 seconds, however, in Ref. 7 Bachelder showed that like the open loop NM frequency it can substantially vary depending on the vehicle dynamics. Reference 7 demonstrated that variability in τ_e is necessary to preserve linearity between phase margin and workload. The output of G_{VE} (β_{ω_c}) is the command signal for the crossover frequency ω_c. The force produced by the operator is exerted on the feel system G_{FS} yielding stick deflection, which is received by the vehicle transfer function G_V and generates the vehicle output m.

Hess' HQM Metric

In Ref. 15 Hess developed a method for predicting Handling Quality Levels (1-3) that would be assigned to a particular vehicle. Hess' proposed Handling Qualities Metric (HQM) computes the maximum magnitude of a transfer function between the inner-loop rate feedback and the command input, normalized by the pilot gain (to remove the effects of control sensitivity). Figure 11 shows the HQMs for various aircraft model transfer functions. Also shown are suggested boundaries between handling-quality levels. These bounds were assigned after reviewing the handling quality ratings assigned to similar controlled elements. Some limitations of this approach include: 1) Task forcing function is not accounted for, so that task cannot be differentiated; 2) Control and display sensitivity effects are not accounted for; and 3) Since the analysis uses a fixed crossover frequency of 2 rad/s, performance may be substantially different from what a pilot might generate or what a poor vehicle configuration might produce. With regard to the third limitation, the HQM approach was not intended to predict performance, but rather to use performance in a comparative manner.

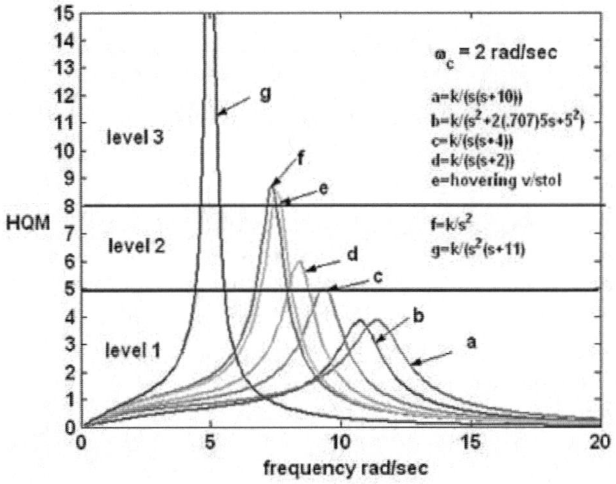

Figure 11. HQM plots for various aircraft models (Ref. 15).

CAPTURING PILOT BEHAVIOR

Each pilot was often observed to execute two different control strategies within the same PACH maneuver. A typical example using actual pilot data is shown in Figure 12a, where the pitch response (shown in blue) in the first and last step change is highly oscillatory, whereas pitch response for the inner two steps is relatively well-behaved. In Figure 12b the stick response associated with the oscillatory pitch is characterized by a sparse pulsing, and the stick response corresponding to the two inner steps consists of a succession of closely-spaced pulses.

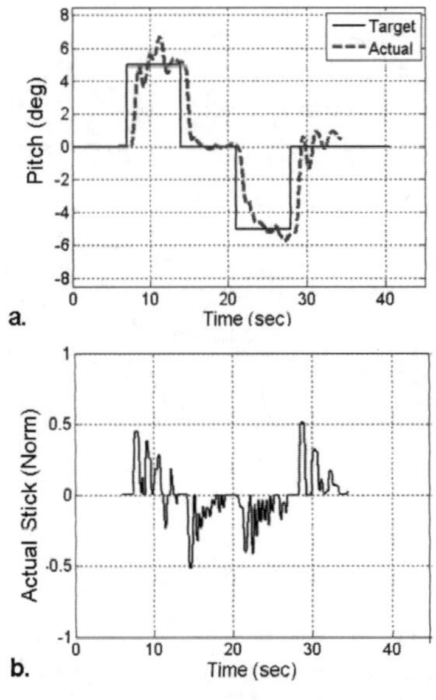

Figure 12. Example pitch-stick control law response: a) Aircraft pitch; b) Pilot stick response.

In Ref. 16 Bachelder developed a pilot modeling technique that employed a cost function that characterized pilot behavior during a pitch sum-of-sines tracking task, incorporating it into the Spare Capacity OPerator Estimator (SCOPE) analysis tool. The pilot cost function assumes the pilot model given by Figure 10 which employs proprioceptive lead/lad feedback. Figure 13a shows the resulting pitch response (magenta line) when SCOPE's proprioceptive equalization is applied to the PACH task. SCOPE's pitch response closely matches the pilot's pitch for the second and third steps.

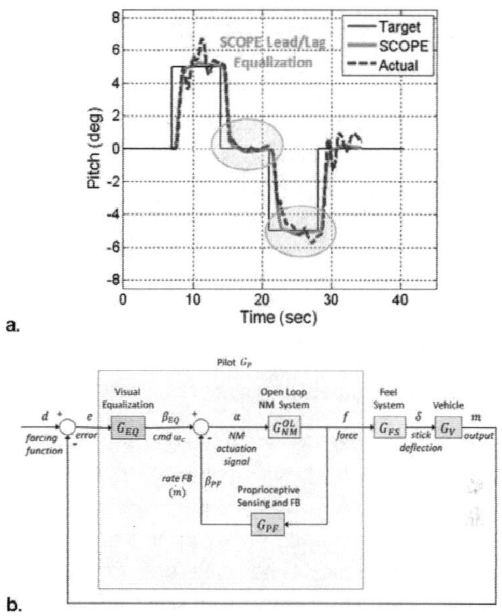

Figure 13. Pitch-stick control law response example using cost function from Ref. 16 : a) Aircraft pitch; b) Pilot model employing proprioceptive feedback.

In Figure 14 the pilot is modeled as a pure gain and time delay that are varied such that the pilot cost function in Ref. 16 as minimized, resulting in pitch oscillations that resemble the actual pilot during the first and last steps.

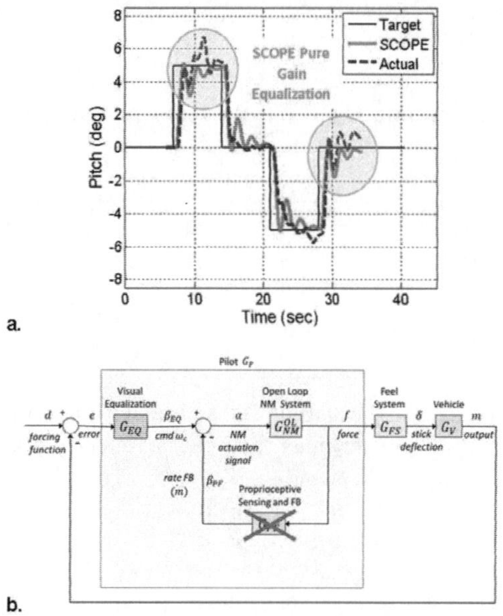

Figure 14. Outer loop pitch-stick control law response example: a) Aircraft pitch; b) Pilot stick response.

In general, it was observed that the poorest interval performance during the PACH maneuver had the greatest influence on pilot HQR rating. Since the pilots usually exhibited gain-only behavior during some portion of most of the PACH runs, the approach employed in this work was to model this worst-case behavior when predicting PACH performance and HQRs.

PILOTED FLIGHT SIMULATION STUDY

In Reference 17 Berger investigated various pitch-axis control laws for high-speed rotorcraft applications using piloted fixed-base simulation. Two Future Vertical Lift-relevant rotorcraft configurations (a lift offset coaxial helicopter with a pusher propeller and a tiltrotor) were employed in the study. This current work examines that study's pitch attitude capture and hold (PCAH) task whose data is compared with the HQRs and performance predicted by SCOPE. The data is taken from two Army experimental test pilots.

Pitch Attitude Capture and Hold Task

A precision tracking task was flown using a display driven by a reference attitude driven by a series of step changes. The stimulus filtering frequency ω_μ at which the workload metric is filtered is 20 rad/s corresponding to approximately the neuromuscular mode frequency. The objectives of this tasks are to evaluate the ability to capture a desired attitude and identify maneuverability limitations, inceptor characteristics, cross coupling, and any PIO tendencies. Step changes were in increments of 5 degrees. Performance criteria for the PACH task is given in Table 1. Two of the criteria for desired performance was that the absolute error between target and actual pitch be less than 1 deg for at the last 5 seconds of each step change, and no PIO tendencies (no more than one overshoot). Adequate performance criterion is less than 2 deg error over the last 5 seconds and no divergent PIO tendencies. Figure 15gives an example Pitch Angle Capture and Hold time history showing desired and adequate performance criteria.

Table 1. Pitch Attitude Capture and Hold performance criteria.

Pitch	Desired	Adequate
• Pitch angle error (from command) tolerance:	±1 deg	±2 deg
• Airspeed deviation tolerance:	±5 kts	±10 kts
• No more than one pitch attitude overshoot on the initial capture of each attitude. Magnitude of overshoot is less than:	1 deg	2 deg
• PIO considerations:	No PIO tendencies	No divergent PIO tendencies
• Inter-axis coupling shall not be	Undesirable	Objectionable

The control laws used in the current study were two inner loop laws (G_{VIL1}, G_{VIL2}) and an outer loop G_{VOL} pitch-to-stick control law given by Eq. 7, 8, and 9.

$$G_{V\_IL1} = \frac{\theta}{\delta} = \frac{K/\tau}{s(s+{}^{1}/_{\tau})} \qquad 7$$

$$G_{V\_IL2} = \frac{\theta}{\delta} = \frac{KT_{\theta_2}\omega^2\left(s+{}^{1}/_{T_{\theta_2}}\right)}{s(s^2+2\zeta\omega s+\omega^2)} \qquad 8$$

$$G_{V\_OL} = \frac{\theta}{\delta} = \frac{KT_{\theta_2}\omega^2/\tau\left(s+{}^{1}/_{T_{\theta_2}}\right)}{s(s+{}^{1}/_{\tau})(s^2+2\zeta\omega s+\omega^2)} \qquad 9$$

In G_{V_IL1}, τ is the pitch rate command model time constant, and for G_{V_OL} it is the flight path rate command model time constant (there is no pitch rate command model in the Outer Loop case). T_{θ_2} is the flight path-attitude lag. The parameter values used for the outer loop control law G_{V_OL} are representative of those for a tilt rotor aircraft and coaxial with a pusherprop helicopter. Pitch attitude bandwidth (ω_{BW_θ}) for the control law configurations ranged from approximately 1.3 to 4.8 rad/s, shown in Figure 16.

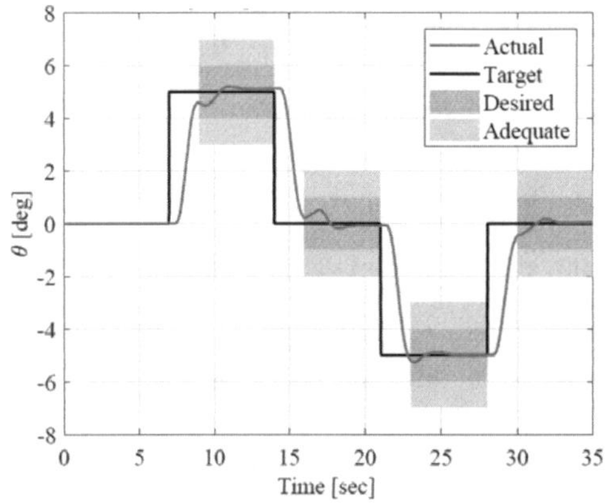

Figure 15. Example Pitch Angle Capture and Hold time history showing desired and adequate performance criteria.

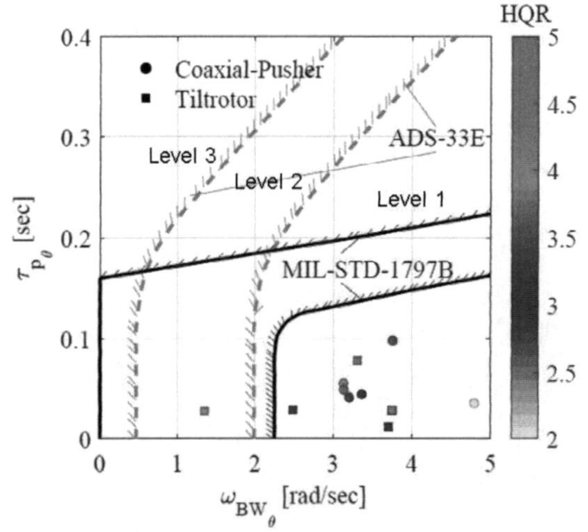

Figure 16. Pitch Attitude Capture and Hold handling qualities ratings plotted on the pitch attitude bandwidth (ω_{BW_θ}) requirement (Ref. 17).

SCOPE HQR AND PERFORMANCE PREDICTION

The SCOPE workload metric can incorporate any number of pilot inputs and outputs, each having its own respective perceptual exponent γ as was demonstrated in Ref. 18. For the PACH task it was determined that in addition to stick rate and error rate, tracking error was a key source of workload. Unlike previous SCOPE-related experimental studies, the minimum and maximum tracking errors observed in the PACH task differed by more than a factor of 10, which made the effect of tracking error on workload prominent. The error, error rate, and stick rate perceptual exponents used were 0.07, 0.2, and 0.2, respectively. To establish the pilot parameters associated with an HQR of 1 for the PCAH task, the MSM pilot model shown in Figure 10 was used with rate command dynamics ($G_V = 1/s$) to generate the baseline workload metric μ^Y, denoted by ${\mu^Y}_{HQR1}$. According to the definition provided by Cooper and Harper's this should correspond to the minimum workload required to produce the desired performance criteria. It was observed that for the PACH task pilot operating point generally lies in the vicinity where minimum RMS tracking error is produced. Figure 17 shows the SCOPE simulation using rate command vehicle dynamics where the pilot is modeled as a direct path gain and a pure time delay (the time delay is itereated to maximize linearity between the open-loop phase margin and the workload metric as described in Ref. 16). Minimum desired performance is specified for the PACH task as an RMS tracking error of less than 1 deg over the final 5 seconds of each pitch step. However, the simulated pilot's operating point (denoted by a red dot in Figure 17a) corresponds to an RMS error of 0.03 deg – much lower than the desired performance criterion. Rarely, if ever, is the evaluation pilot asked to 'thottle back' on intensity of effort to what is needed to produce minimum desired performance so that he/she is generating the workload associated with that minimum level of performance. Instead, pilots tend to operate at a state that balances workload with performance when conducting flight evaluation – even if this generates performance that is considerably greater than what is required. The maximum tracking tracking observed in the thirteen PACH evaluation runs conducted in Ref. 17 was 0.74 deg, and the combined average RMS error was 0.42 deg. Again, only 1 deg was required to meet the desired performance. SCOPE computes the baseline ${\mu^Y}_{HQR1}$ at the pilot's operating point which is at the crossover frequency and workload metric corresponding to the minimum RMS tracking error as shown in Figure 17a and b. This workload metric is shown in the first row of the table seen in Figure 18, where the highlighted baseline ${\mu^Y}_{HQR1}$ is 0.933.

The workload metric ${\mu^Y}_{HQR6}$ is computed for HQR6, which corresponds to the minimum adequate performance (2 deg RMS tracking error) at maximum tolerable workload. Here the aircraft dynamics are made progressively more difficult by decreasing the root λ in Eq. 10 until the maximum possible performance coincides with adequate performance.

$$G_V = \frac{\lambda}{s(s+\lambda)}, \text{decreasing } \lambda \qquad 10$$

The parameters associated with HQR6 are seen in the sixth row of the table seen in Figure 18, where ${\mu^Y}_{HQR6}$ is computed at 1.1478 (highlighted). The dynamics that generated this were $G_V = \frac{0.42}{s(s+0.42)}$ (essentially acceleration command dynamics).

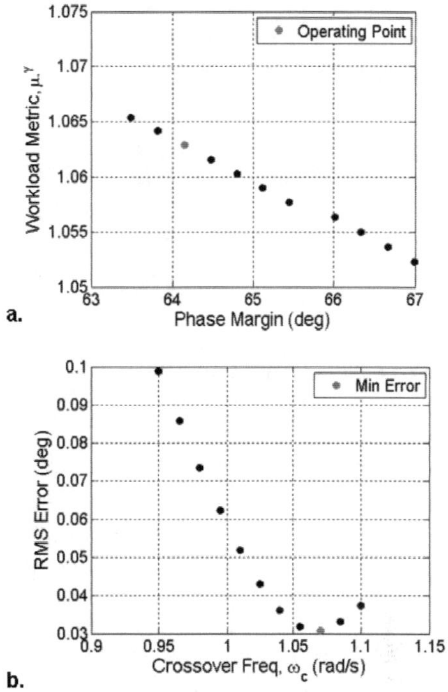

Figure 17. SCOPE simulation using rate command: a) Tracking performance versus crossover frequency; b) Workload metric versus phase margin.

Aircraft Characteristics	Demands on the Pilot in Selected Task	Pilot Rating	G_V	Workload Metric $\mu^Y = (\sigma_e \sigma_{\dot{e}} \sigma_{\dot{\delta}})^Y$	Perf	PM (deg)
Excellent, highly desirable	Compensation not a factor for desired performance	1	$\frac{1}{s}$	0.9330	Des	64
Good, negligible deficiencies	Compensation not a factor for desired performance	2				
Fair – some mildly, unpleasant deficiencies	Desired performance requires minimal compensation	3				
Minor but annoying deficiencies	Desired performance requires moderate compensation	4				
Moderately objectionable deficiencies	Adequate performance requires considerable compensation	5				
Very objectionable but tolerable deficiencies	Adequate performance requires extensive compensation	6	$\frac{0.42}{s(s+0.42)}$	1.1478	Adeq	31
Major deficiencies	Adequate performance not attainable with maximum tolerable compensation. Controllability is not in question	7				
Major deficiencies	Control requires considerable compensation	8				
Major deficiencies	Control requires intensive concentration	9				
Major deficiencies	Control will be lost during some portion of the required operation	10				

Figure 18. HQR workload metric references for PCAH task.

The workload metrics for HQR2-5 are linearly interpolated between ${\mu^Y}_{HQR1}$ and ${\mu^Y}_{HQR6}$, shown in Figure 19. The phrasing associated with "Adequacy for Selected Task" adjacent to HQR7-9 refers to tolerable workload, which implies that workload for HQR1-6 is considered tolerable. Thus the only difference then between HQR6 and HQR7 is performance (the former being adequate and the latter inadequate), and they are accordingly both upper-bounded by the same μ^Y of 1.148. The textual compensation ascribed to HQR8 is precisely the same used for HQR5 (considerable), so the bounds on μ^Y are made identical (the two HQRs differ in performance and PM criteria). HQR9 denotes concentration as 'intensive', which is assumed to be on par with 'extensive compensation' annoted for HQR6, and ${\mu^Y}_{HQR9}$ is assigned to

Aircraft Characteristics	Demands on the Pilot in Selected Task	Pilot Rating	Workload Metric $\mu^Y = (\sigma_e \sigma_{\dot{e}} \sigma_\delta)^Y$	Perf	PM (deg)
Excellent, highly desirable	Compensation not a factor for desired performance	1	<=0.933	>Des	>20
Good, negligible deficiencies	Compensation not a factor for desired performance	2	>0.933 <=0.976	>Des	>20
Fair – some mildly, unpleasant deficiencies	Desired performance requires minimal compensation	3	>0.976 <= 1.019	>Des	>20
Minor but annoying deficiencies	Desired performance requires moderate compensation	4	>1.019 <= 1.062	>Des	>20
Moderately objectionable deficiencies	Adequate performance requires considerable compensation	5	> 1.062 <= 1.105	>Adeq	>20
Very objectionable but tolerable deficiencies	Adequate performance requires extensive compensation	6	> 1.105<=1.148	>Adeq	>20
Major deficiencies	Adequate performance not attainable with maximum tolerable compensation. Controllability is not in question	7	<=1.148	<Adeq	>20
Major deficiencies	Control requires considerable compensation	8	> 1.062 <= 1.105	<Adeq	>5, <20
Major deficiencies	Control requires intensive concentration	9	> 1.105<=1.148	<Adeq	>5, <20
Major deficiencies	Control will be lost during some portion of the required operation	10		<Adeq	<5

Figure 19. Completed HQR workload metric matrix for PCAH task (reference values highlighted).

the bounds of $\mu^Y{}_{HQR6}$. Since control is lost for HQR10 the criteria assigned to this are performance (inadequate) and PM (< 5 deg).

Finally, the pitch dynamics used in the actual task (Eq. 9) are employed in the pilot model and crossover frequency iterated to establish the pilot's operating point. The completed matrix in Figure 19 will be used to compute the ordinal HQR estimates for the evaluation task configurations. Continuous HQR estimates are also computed using the slope and bias produced by the line formed by $\mu^Y{}_{HQR1}$ and $\mu^Y{}_{HQR6}$.

RESULTS

Bedford ratings were not collected from the pilots for the PACH task in Ref. 17 Continuous HQRs were computed using the slope and bias produced by the line formed by $\mu^Y{}_{HQR1}$ and $\mu^Y{}_{HQR6}$. Figure 20 compares the predicted SCOPE HQRs with actual, where the matching coefficient R_{mod} is 0.94. The regions associated with Levels 1and 2 are shown, with SCOPE predicting the correct Level for all but one of the thirteen configurations- the one incorrect estimate is off by less than a quarter of a rating.

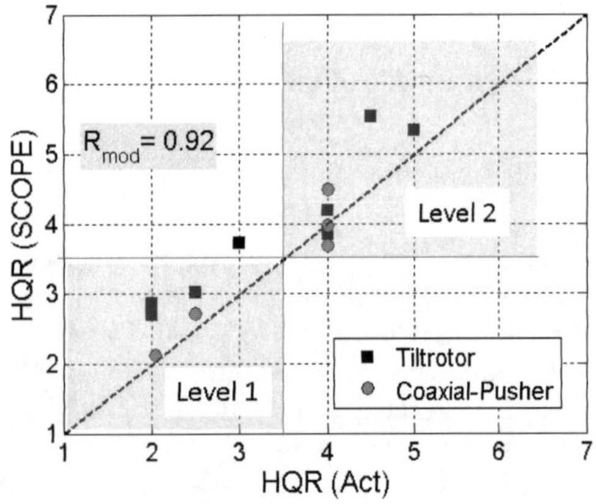

Figure 20. Predicted SCOPE HQR versus actual showing Level 1 and 2 regions.

Figure 21a plots the actual HQRs on the ADS-33 pitch attitude bandwidth requirement, where ADS-33 predicted only one out of the seven Level 2 configurations correctly.

Figure 21b shows the two highest HQRs were well-estimated by SCOPE, but lie solidly within the ADS-33 Level 1 region.

Figure 22 shows an updated boundary proposed by Ref. 17 for the ADS-33 pitch attitude dropback specification based on the HQRs collected for the PACH task. SCOPE can be used to map high-fidelity HQ boundaries by systematically generating predicted HQRs for any pair of metrics and/or parameters.

a.

b.

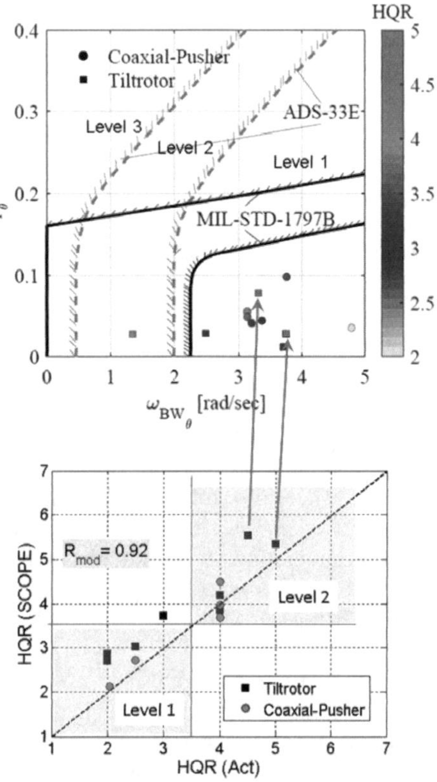

Figure 21. a) Actual HQRs plotted on the ADS-33 pitch attitude bandwidth requirement; b) Predicted SCOPE HQR versus actual showing highest two ratings relative to the ADS-33 bandwidth requirement.

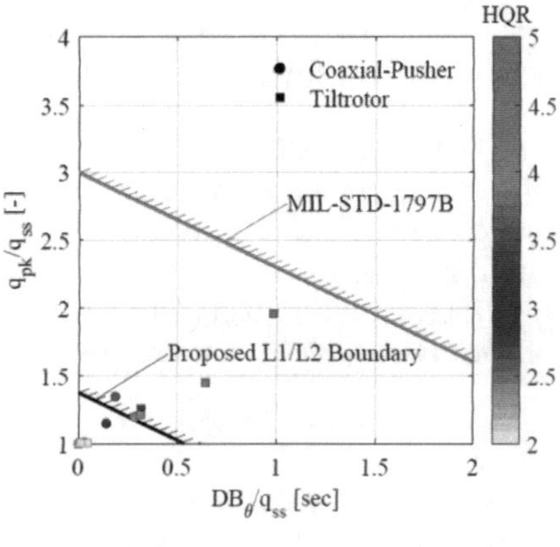

Figure 22. Pitch Attitude Capture and Hold handling qualities ratings plotted on the pitch attitude dropback requirement with Level 1/Level 2 boundary proposed by Ref. 17.

The Pearson Correlation Coefficient R measure is a metric characterizing how well observed outcomes match a best-fit line through the data. The interest here is to measure the agreement between an observation (Y_{obs}) and an estimate (Y_{est}) of that observation' (if agreement was perfect all the points would lie on the line y = x). The R metric quantifies the degree of any linear correlation between Y_{obs} and Y_{est}, whereas the goodness-of-agreement sought should only consider one specific linear correlation based on $Y_{est} = Y_{obs}$. For this, R is modified as shown in Eq. 11

$$R_{mod} = \sqrt{1 - \frac{\sum(X-Y)^2}{\sum X^2}} \quad 11$$

Figure 23 compares the predicted SCOPE tracking performance with actual, where the matching coefficient R_{mod} is 0.85. Predicted SCOPE crossover frequency is compared with actual in Figure 24 yielding a matching coefficient of 0.88.

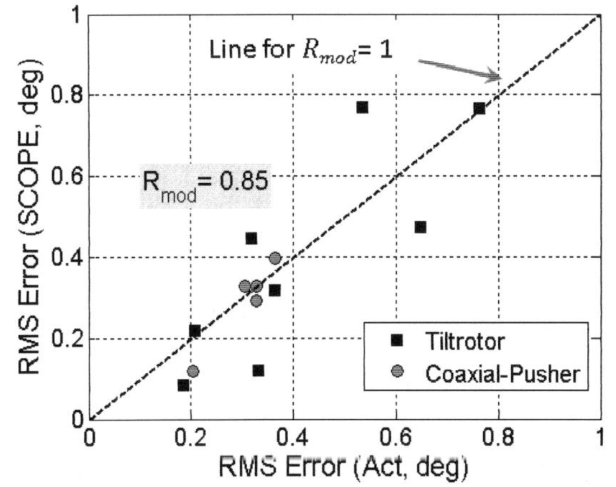

Figure 23. Predicted SCOPE RMS error versus actual.

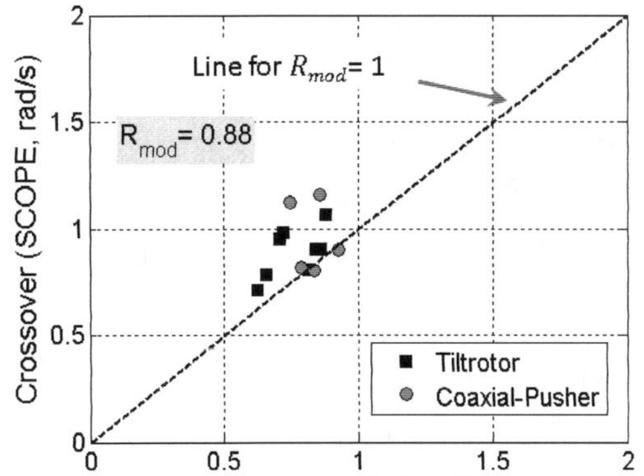

Figure 24. Predicted SCOPE crossover versus actual.

DISCUSSION

HQR performance criteria specification

Cooper and Harper divide the total workload into two categories for handling qualities rating: 1) Best-case task-execution workload (i.e., using rate command), and 2) Workload compensating for aircraft deficiencies (hence the term *compensation*), which is the workload in excess of that required for best-case task-execution. Any handling qualities task using the best-case configuration should therefore receive a rating of HQR1 if the desired performance criteria are met. When generating the baseline HQR1 not only should the vehicle dynamics be ideal (i.e., rate command), but the inceptor and display elements should also have optimum characteristics. For instance, in Ref. 9 it was observed that tracking performance was significantly poorer during rate command control if the inceptor's spring-centering force was removed. This raises an important consideration - if the task maneuver and forcing function are such that desired performance cannot be attained using the best-case configuration, then HQR5 is the most favorable rating possible (assuming adequate performance is achievable using rate command). As it is inconsistent to specify criteria that cannot be met under ideal conditions, it is proposed that desired performance criteria should not exceed the best-case performance.

Tasks such as shipboard hover can have particularly challenging forcing functions due to wind turbulence around the ship superstructure. Helicopter flight controls designed for shipboard operation such as the MH-60R Seahawk have gust alleviation capability which provides substantial stabilization in the presence of high frequency forceful gusts. When such stabilization exists, it should be included with the idealized command dynamics to generate the baseline workload metrics $\mu^Y{}_{HQR1}$ and $\mu^Y{}_{HQR6}$.

Most flight tasks require multi-axis operation, and there is often control coupling between axes. Spare capacity and performance are reduced from attending to multiple axes of control compared to single-axis control (Ref. 10). Pertaining to multi-axis HQR tasks, performance criteria and baseline workload metrics for each axis of control should be based on the best-case assumption of single-axis, rate-commanded dynamics. In this way workload arising from phenomena such as cross-coupling appears as compensation for aircraft deficiencies.

Assignment of HQR baseline compensation and phase margin

It could be assumed the difference in compensation between HQR1 and 2 was too subtle to express effectively in words (the HQR scale describes compensation identically for the two), but a numerical compensation gate captures the difference. The most concrete points for computing reference compensation are HQR1 and HQR6; the former when desirable performance is first attained using rate command dynamics, and the latter when adequate performance is last encountered at maximum workload (and controllability is still assured). For simplicity the workload metric for HQR2-5 is linearly interpolated between these two anchor points.

In Reference 1, a PM of 45 degrees is assigned to Level I (HQR = 1 through 3), assuming a pure gain pilot is closing the loop around the inceptor and vehicle. Reference 19 allows for 50% degradation for worst-case scenario (i.e., Level 2), or

22.5 degrees. In this work, the strategy adopted for mapping PM and quantitative compensation to HQRs was to make the numerical gates as consistent with the descriptors as possible. Mention of controllability starts at HQR8; based on pilot data (Refs. 20 and 9) problems related to stability were observed at phase margins of roughly 20 degrees. PM will improve moving up the scale towards HQR1, and it probably increases somewhat linearly. But the critical discriminator (as evidenced by the HQR phrasing) is compensation and PM will follow. Imposing complex exit criteria through more PM tiers would make the technique stiff and liable to 'blind spots', hence a single PM floor of 20 degrees was used as the PM criterion for HQR1-7.

Summary of SCOPE HQR and Bedford estimation technique

SCOPE's HQR estimation is unique to a given task, which is defined by maneuver, disturbance, target states, and performance criteria. The SCOPE HQR compensation matrix will then apply to any set of vehicle dynamics. If display or inceptor dynamics are to be evaluated, they must be varied separately to avoid confounding effects (applying SCOPE HQR estimation to displays and inceptors will be addressed in future work).

The steps needed to construct a SCOPE HQR prediction matrix is summarized as follows:

1) Establish the two workload metric baselines (HQR1 and HQR6). From the set of optimally generated PM, performance, and $G_{\delta d}$ responses the pilot operating point was selected.

2) Interpolate compensation between HQR1 and HQR6 for HQR2-5. As shown in Figure 19, HQR7-9 re-uses the workload metric data for HQR5-6 based on near-matching descriptions of compensation.

A similar process constructs the Bedford rating prediction matrix, except that proportional command dynamics vice rate command (Ref. 16) is used to generate the Bedford1 workload metric. Additionally, Beford7-10 extrapolate from the baseline values Bedford1 and Bedford6.

SUMMARY AND CONCLUSIONS

The approach presented in this work estimates workload and HQRs by establishing baseline workload levels for a given task using reference vehicle command dynamics. These baseline levels are generated by the Spare Capacity OPerator Estimator (SCOPE), which estimates workload with a model of pilot neuromuscular feedback (the Modified Structural Model) and a cost function that captures pilot behavior. Using the baseline workload levels and the original definitions found in the HQR scale, quantitative values from SCOPE are mapped to the HQR descriptors 'compensation' and 'controllability'. SCOPE's performance and HQR estimates for a piloted simulation conducting a pitch attitude capture and hold task compared favorably with the performance and HQRs collected. SCOPE did not employ pilot data for calibration. Pilot HQR for the PACH task appears to be based on the pilot's response when acting as a pure gain in the direct path with little or no proprioceptive feedback. SCOPE's performance and crossover predictions matched well with the actual values, indicating that the pilot operating point lies in the vicinity where minimum RMS tracking error is produced (for the PACH task).

The central contributions of this work are:

1. A concise definition and quantitative metric of pilot compensation is developed.

2. The proposed pilot behavioral model appears to capture the primary drivers of performance and HQR during the PACH task.

3. SCOPE's methodology using performance prediction and workload baselining will allow designers to predict task performance and provide diagnostic and prognostic workload and HQR capability.

4. SCOPE's workload metric accounts for control and display sensitivity effects on workload, HQR, and performance, enhancing operational utility.

5. SCOPE can be used to ensure the performance criteria for a handling qualities evaluation task are harmonized with the task.

FUTURE WORK

SCOPE will be applied to other handling qualities evaluation tasks such as MTEs requiring large initial inputs followed by terminal tracking (break turn, high speed acceleration, high speed deceleration). The effects of pulsive control and limb motor noise (Ref. 21) will be integrated into SCOPE.

It will also be demonstrated how SCOPE can be applied to UAS design. A UAS whose flight controller mimics a skilled human pilot and whose open loop response (controller-vehicle suite) produces favorable human-in-loop HQRs would have reduced risk of adverse outcomes if the human assumes control of flight from automation. Additionally, a UAS design requiring reduced 'workload' should see improvements in lifetime reliability, weight, and overall power requirements.

Author contact:
Edward Bachelder edward.bachelder@sjsu.edu
Bimal Aponso Bimal.L.Aponso@faa.gov

ACKNOWLEDGMENTS

This work was supported by cooperative agreement 80NSSC19M0228 between the U.S. Army DEVCOM Aviation & Missile Center and San Jose State University.

REFERENCES

1 Anon., "Handling Qualities Requirements for Military Rotorcraft," Aeronautical Design Standard—33 (ADS-33E-PRF), US Army Aviation and Missile Command, Mar. 2000.
2 Mitchell, David, and Bimal Aponso. "Reassessment and extensions of pilot ratings with new data." 17th Atmospheric Flight Mechanics Conference, 1990.
3 Hoh, Roger. "Lessons learned concerning the interpretation of subjective handling qualities pilot rating data." 17th Atmospheric Flight Mechanics Conference. 1990.
4 Memon, W. A., White, M. D., Padfield, G. D., Cameron, N., & Lu, L. (2021). Helicopter handling qualities: A study in pilot control compensation. The Aeronautical Journal, 1-35.
5 Cooper, George E., and Robert P. Harper. The use of pilot rating in the evaluation of aircraft handling qualities. National Aeronautics and Space Administration, 1969.
6 Bachelder, Edward N., and Martine Godfroy-Cooper. "Pilot Workload Estimation: Synthesis of Spectral Requirements Analysis and Weber's Law." AIAA Scitech 2019 Forum. 2019.
7 Bachelder, Edward, and Bimal Aponso. "Theoretical Framework Unifying Handling Qualities, Workload, Stability, and Control." Vertical Flight Society 77th Annual Forum & Technology Display, Virtual. 2021.
8 Bachelder, E., Lusardi, J., and Aponso, B. "Neuromuscular Response Comparison for Center and Side Stick Positions." Vertical Flight Society 78th Annual Forum & Technology Display, Fort Worth, TX. 2021.
9 Bachelder, Edward N., and Bimal L. Aponso. "Human Pilot Control Adaptation: A Physiological Interpretation." AIAA Scitech 2022 Forum, San Diego, CA. 2022.
10 David G. Mitchel, Bimal L. Aponso, and Roger H. Hoh. Minimum Flying Qualities. Volume 1. Piloted Simulation Evaluation of Multiple Axis Flying Qualities. USAF WRDC-TR-89 3125, Jan. 1990.
11 Mitchell, David G, Bimal L Aponso, And Mark J Detroit, "Simulation Investigation of Multiple Axis Flying Qualities." AIAA Guidance, Navigation and Control Conference, Portland, OR. 1990.
12 Hess, Ronald A. "Unified theory for aircraft handling qualities and adverse aircraft-pilot coupling." Journal of Guidance, Control, and Dynamics 20.6 (1997): 1141-1148.
13 Bachelder, Edward N., and Bimal L. Aponso. "Handling Qualities and Aircraft Pilot Coupling: New Evidence of Neuromuscular Origins." AIAA Scitech 2022 Forum, San Diego, CA. 2022.
14 McRuer, D. T. and Krendel, E. S., "Mathematical Models of Human Pilot Behavior," No. AGARDograph No. 188, November, 1973.
15 Hess, Ronald A. "Simplified approach for modelling pilot pursuit control behaviour in multi-loop flight control tasks." Proceedings of the Institution of Mechanical Engineers, Part G: Journal of Aerospace Engineering 220.2 (2006): 85-102.
16 Bachelder, Edward, Bimal Aponso, Tom Berger. "A Theoretical Basis for Predicting Pilot Performance, Workload, and Handling Qualities." *AIAA Scitech 2023 Forum*. 2023.
17 Berger, Tom, Mark B. Tischler, and Joseph F. Horn. "High-Speed Rotorcraft Pitch Axis Response Type Investigation." Proceedings of the Vertical Flight Society 77th Annual Forum, online. 2021.
18 Bachelder, Edward, et al. "Pilot Workload and Performance Assessment for a Coaxial-Compound Helicopter and Tiltrotor During Aggressive Approach." Vertical Flight Society 77th Annual Forum & Technology Display, Virtual. 2021.
19 Society of Automotive Engineers, Aerospace Standard AS94900, Flight Control Systems, SAE International, 2007.
20 Bachelder, Edward, and Bimal Aponso. "A Theoretical Framework Unifying Handling Qualities, Workload, Stability, and Control." Vertical Flight Society 77th Annual Forum & Technology Display, Virtual. 2021.
21 Bachelder, E., Lusardi, J., and Aponso, B. "Interrelationship Between Limb Motor Noise, Limb Motion and Tracking Performance," Vertical Flight Society 79th Annual Forum & Technology Display, West Palm Beach, FL, 2023 (In Press).

Handling Qualities Evaluations of Active Inceptors with Varying Force-Feel Characteristics

Michael Jones
Director of Engineering Services
Systems Technology, Inc.
Hawthorne, CA, USA

David Klyde
Technical Fellow
Systems Technology, Inc.
Hawthorne, CA, USA

Dakota Musso
Resident Engineer
Systems Technology, Inc.
Hawthorne, CA, USA

Cody Fegely
HQ Principal Engineer
Bell Textron Inc.
Fort Worth, TX, USA

Luke Gillett
FCS Staff Engineer
Bell Textron Inc.
Fort Worth, TX, USA

Seung Baik
HQ Principal Engineer
Bell Textron Inc.
Fort Worth, TX, USA

ABSTRACT

Active Inceptors in fly-by-wire aircraft offer new possibilities for control law reconfigurability, haptic cueing and system redundancies. Based on recent active inceptor research, new requirements for force-feel characteristics are proposed. Results are obtained from a simulation campaign investigating the influence of dynamic force-feel characteristics on the overall air-vehicle handling qualities (HQs). For this study, a rate-command type controller is evaluated in a pilot-in-the-loop simulation facility. Five mission task elements (MTEs) are flown with a range of force-feel characteristics (stick damping and stick natural frequency). Pilot comments on the active inceptors indicate clear differences and preferences between inceptor configurations. Generally, inceptor configurations with lower damping and higher natural frequency are preferred. Some minor differences, with respect to the MTEs flown, were noted and these may be explored in more detail in future research.

INTRODUCTION

The use of Active Inceptors can improve rotorcraft handling qualities (HQs) to reflect specific operating conditions or specific operational modes. Active Inceptors replace mechanically linked pilot/co-pilot controls, to allow variations in force-feel parameters during real-time operations. Active inceptors offer additional benefits, such as electronic control axis de-coupling and haptic feedback. Previous research has shown various benefits of active inceptors, through both simulation and flight test campaigns [1], [2], [3] [4]. This paper presents results from a new simulation campaign to determine changes in HQ ratings due to active inceptor sidestick force-feel parameters. The study includes a range of mission tasks and flight speeds. The flight conditions flown, the inceptor configurations tested, the tasks selected, and the results are discussed.

A number of studies have investigated the optimal settings for both static and dynamic force-feel settings using active inceptors. Efforts by researchers at the U.S. Army Capabilities Development Command (DEVCOM) Aviation & Missile Center (AvMC) [1] and the German Aerospace Center (DLR) [5] are notable. The former conducted tests using an active center stick, the latter using an active inceptor sidestick. From these studies, boundaries have been drawn for the dynamic force-feel characteristics of active inceptors, including rate command and attitude command flight control strategies. These boundaries, which define the damping and natural frequency of the inceptor force-feel system, are contained in proposed revisions to Aeronautical Design Standard – 33E [6] and are shown below. Figure 1 shows the boundaries for attitude command systems, while Figure 2 shows rate command boundaries of primary interest for this investigation. These boundaries have been drawn predominantly from tests conducted using the center stick active inceptor. Comparing Figure 1 and Figure 2, the Level 1 region is most stringent, requiring the narrowest range of damping and natural frequency, for the rate command system. Generally, as the damping and natural frequency increase, the

Presented at the VFS Handling Qualities Technical Meeting, Huntsville, Alabama, USA Feb. 7-8, 2023.

This research was partially funded by the Government under Agreement No. W911W6-19-9-0002.

HQs improve. For both rate and attitude command systems, cases with high natural frequency and low damping are considered Level 3. This is due to biodynamic feedback effects which may occur for this configuration. Typically, all cases greater than HQL 2 are critically damped or overdamped. From studies conducted at DLR using active inceptor sidesticks, it appears that the natural damping of the pilot's arm may lead to a reduction in damping required for Level 1 HQ, with respect to the boundaries shown in Figure 2 [5].

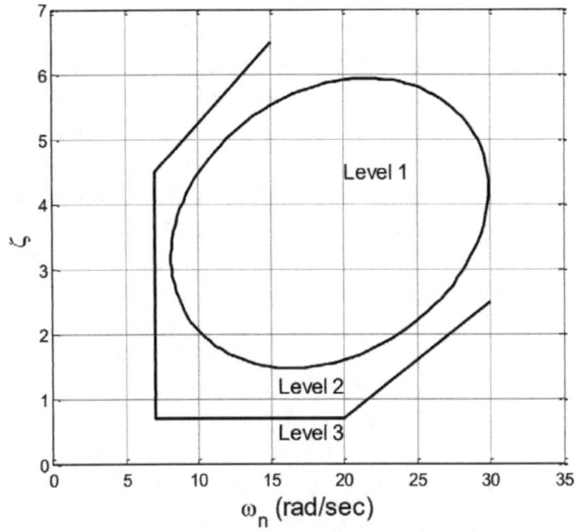

Figure 1: Proposed Boundaries for Active Inceptors, Attitude Command Response

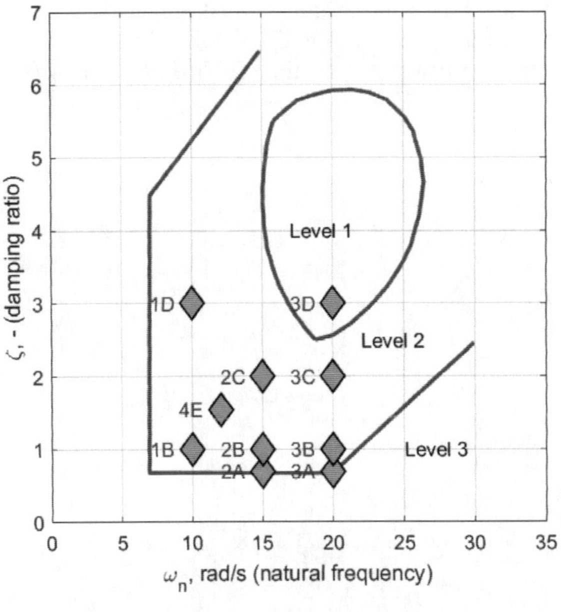

Figure 2: Proposed Boundaries for Active Inceptors, Rate Command Response, and Test Points

TEST CAMPAIGN SET-UP

Over a two-week period, formal force-feel inceptor evaluations were conducted using a fixed-base simulation facility at Bell Textron Inc., Fort Worth, TX. Three test pilots participated in conducting approximately 20-25 hours of simulation testing. Throughout the test campaign, changes to the sidestick force-feel characteristics were the only changes made. The vehicle model and control system parameters were left unchanged, so as to not introduce additional variables.

Experimental Setup

The inceptor force-feel evaluations were conducted in is a fixed-based simulator. It featured a single pilot station with a multi-functional display, a right-hand sidestick, a left-hand collective, and pedals. The simulation graphics were projected onto a 3D high-resolution domed display that provided a +180 deg field-of-view. The simulation featured multiple environments and included the virtual courses required for conducting the various selected evaluation tasks.

For the experiment, both left- and right-hand sidesticks were initially evaluated. The left-hand sidestick was configured to have functionality similar to a traditional collective control inceptor, as done in previous investigations (e.g., DLR research). The right-hand sidestick was configured to control longitudinal and lateral motion. Due to the selected mission task elements (MTEs) and the availability of upper-modes (i.e., height-hold), the pilot was rarely required to perform corrections for height deviation and therefore, the left-hand sidestick was not a factor in the investigation. Thus, for purposed of this evaluation, all changes were made purely in the right-hand inceptor.

Figure 3 shows the static force-feel characteristics used for the investigation. These characteristics were fixed during the investigation to limit the variables while allowing the focus to be on dynamic characteristics. Figure 3 shows a slight difference in positive and negative static friction to account for the pilot's arm position.

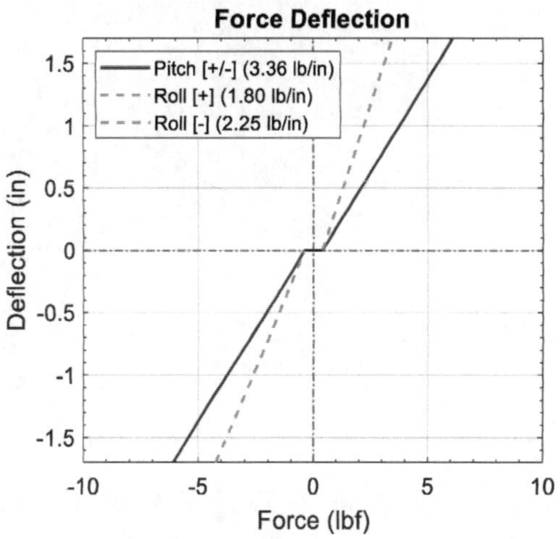

Figure 3: Force-Feel Static Inceptor Characteristics

To investigate the control law configurations with more stringent requirements, a baseline rate command control system was used to select the control inceptor damping/natural frequency boundaries. It is anticipated that a rate command type of controller would be most affected by changes to dynamic force-feel configurations.

A design space containing 10 possible right-hand control inceptor dynamic configurations, was developed to account for configurations with damping values up to 3 and natural frequencies to 20 rad/s. The natural frequency was limited by the available range of the hardware at the time of the experiment. All configurations considered are shown in Figure 2.

Referring to Figure 2, two underdamped cases were selected in the Level 3 region. In previous DLR investigations, results showed lower damping requirements for sidestick inceptors due to the armrest and inceptor position [5]. Therefore, the underdamped cases were tested to see whether these may be suitable for sidestick inceptors. Prior to the investigation it was noted that these configurations may be susceptible to biodynamic feedback, particularly in the real aircraft, where aircraft accelerations and vibration are at play. Bio-dynamic feedback was found to impact certain results in previous studies, particularly with higher natural frequency cases.

The three pilots who participated in the campaign (Pilots A, B, C) were all experimental test pilots, familiar with the HQ rating scales and approach used in the campaign. Pilot C had significant experience flying the physical aircraft being modeled.

Test Matrix

Test evaluations included Pilot feedback consisting of general comments, questionnaires, and formal qualitative ratings. The Cooper-Harper scale (HQRs) [7] served as the basis for handling qualities ratings. Additional feedback regarding workload and PIO tendencies were collected using the Bedford Workload Rating scale (BWL) [8] and the Pilot-Induced Oscillation Tendency Rating scale (PIOR) [10] respectively. Pilots were also asked to complete the questionnaire shown in the Appendix.

Five mission task elements (MTEs) were selected to evaluate both hover and forward-flight conditions. Two of these MTEs (Hover and Slalom) were taken directly from ADS-33E-PRF [9] and used the Scout/Attack tolerances. These two MTEs were selected on the basis that they had previously been used in many of the active inceptor studies undertaken by the US Army and DLR.

An additional Roll Step MTE was selected as an alternative to the Slalom MTE. An issue with the Slalom MTE is its lack of tight, high gain, closed-loop control. The Roll Step MTE was proposed by Meyer and Padfield in Ref [11] because this MTE incorporates both large changes in roll attitude and tracking elements. The task aggression is driven by the flight speed. The course position remains fixed for different flight speeds and therefore the required roll attitudes and time allowed to complete runway transitions become larger and shorter respectively as forward flight speed increases. For this investigation, the Roll Step MTE was flown at 120kts. This was found to exhibit an appropriate level of aggression for the vehicle tested. Figure 4 shows the Roll Step course used during this investigation. Figure 5 a) shows the whisker symbology provided to the pilot display for the Sum-of-Sines tracking task. The objective for each commanded attitude is to maintain the orange dot capture and hold within the green circles for pitch. Simultaneously, the pilot captures and holds the green line within the diagonal whisker bounds for roll. Figure 5 b) shows a command signal for the Sum-of-Sines tracking task with the scoring time limited to a 60 second window.

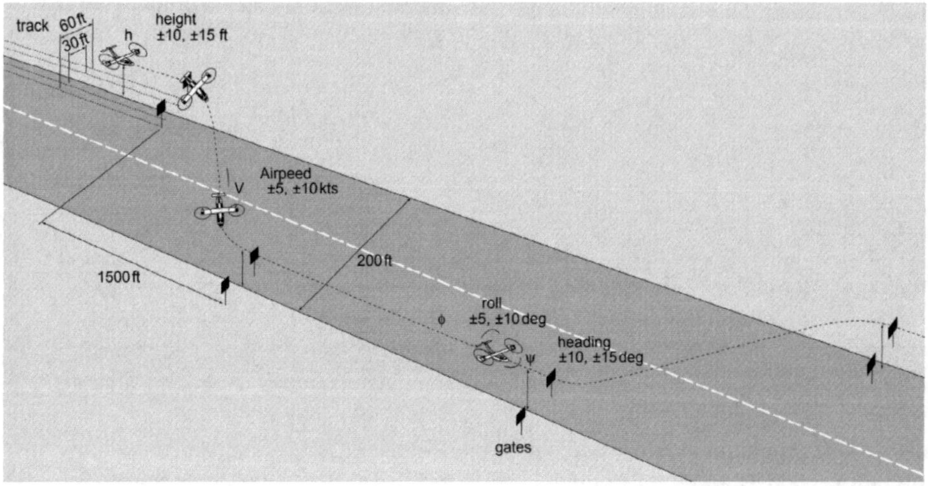

Figure 4: Suggested Course for the Roll Step (Reproduced from Reference [11])

In addition, both independent roll and pitch sum-of-sines tracking tasks were used as single axis tasks, predominantly to determine the pilot characteristics. These tasks were developed during Vertical Lift Consortium (VLC) simulation campaigns [12]. Piloting tasks and target forward flight speeds are shown in Table 1, additional targets and tolerances can be required, specific by MTE. Task performance was assessed by pilot feedback and associated discussion and included qualitative evaluation of performance during the completion of test points. The assessment also included objective qualitative performance analysis from the recorded flight data. Due to time constraints, not all inceptor configurations were flown for all MTEs. The inceptor force-feel configurations selected for each MTE are shown in Table 2, indicating that the majority of configurations were well represented.

Table **3** shows the performance tolerance for the Sum-of-Sines task used during the evaluations. More details can be found in Reference [12].

Table 1: Selected MTE Tasks and Associated Forward Flight Speed

	Target Airspeed (KTAS)		
Task	0-10	20-60	120-150
Hover MTE	X		
Slalom MTE		60	
SoS Tracking Roll			120
SoS Tracking Pitch			120
Roll Step MTE			120

Table 2: Inceptor Configurations Flown with Respect to MTE

MTE	Force-Feel Configurations
Hover	1B, 1D, 2A, 2C, 3B, 3D, 4E
Roll Step	1B, 1D, 2A, 2C, 3B, 3D, 4E
Slalom	1B, 1D, 2A, 2C, 3A, 3B, 3C, 3D, 4E
SoS Pitch	1B, 1D, 2A, 2B, 2C, 3B, 3C, 3D, 4E
SoS Roll	1B, 1D, 2A, 2C, 3B, 3D, 4E

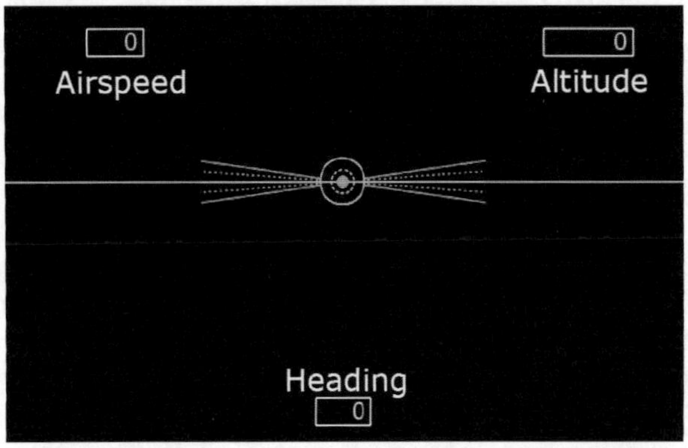

a) Cockpit Display During Sum-of-Sines Task

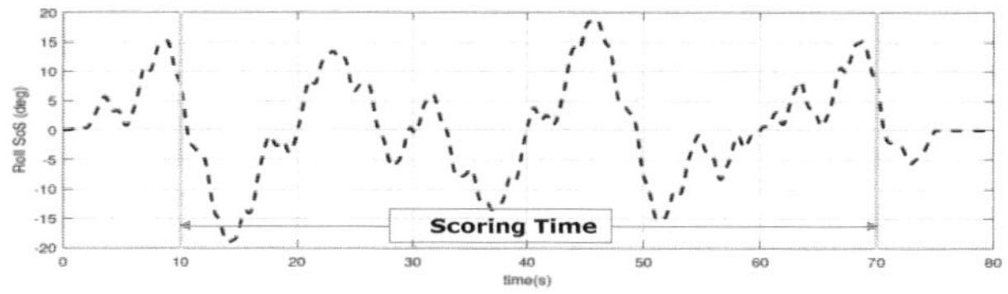

b) Sum-of-Sines Command Signal

Figure 5: Sum-of-Sines Task References [12]

Table 3: Performance Tolerances for Roll and Pitch Tracking

	Desired	Adequate
Pitch: at least X% of the scoring time within pitch attitude error tolerance:	50% ±1°	75% ±2°
Roll: at least X% of the scoring time within roll attitude error tolerance:	50% ±5°	75% ±10°
PIO Considerations:	No PIO tendencies	No divergent PIO tendencies
Inter-axis coupling shall not be	Undesirable	Objectionable

RESULTS

This section shows selected results obtained during the investigation. All ratings obtained during the investigation are given in the Appendix.

Subjective Results for the Evaluation of Sidestick Control Inceptors

Pilots flew the identified inceptor configurations over the course of several days, during which the pilots flew predetermined MTEs with configurations preset. After a brief series of practice runs, to gain familiarity with each MTE/feel-system configuration, the pilots performed several scoring runs. Following these runs, the pilots gave a formal rating using the questionnaire and HQ, BWL, and PIO rating scales. Overall, the pilots awarded HQRs within the Level 1 and Level 2 regions. Overall, the Sum-of-sines pitch tracking task was found to exhibit lower HQ ratings but remained at Level 2 throughout.

A good example of the HQR results (with respect to various MTEs) is shown in Figure 6 and Figure 7 where the respective HQ and Bedford workload ratings are shown for the 4E configuration. The numbers in the plot (n#) identify the number of times a rating was awarded. As shown, some of the cases were only flown twice. The Sum of Sines pitch task shows a large spread in the rating, particularly regarding the Bedford Workload Rating results.

Hover MTE

The Hover MTE is a multi-axis maneuver that is initiated at a ground speed between 6 and 10 knots and at an altitude less than 20 ft. The target hover point is oriented approximately 45 degrees relative to the initial heading of the rotorcraft. The target hover point is a repeatable, ground-referenced point from which rotorcraft deviations are measured. The ground track should be such that the rotorcraft will arrive over the target hover point.

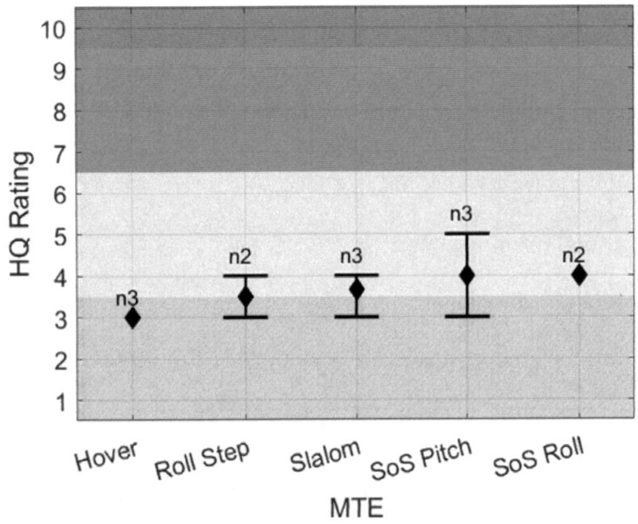

Figure 6: Summary of HQRs Obtained, Configuration: 4E

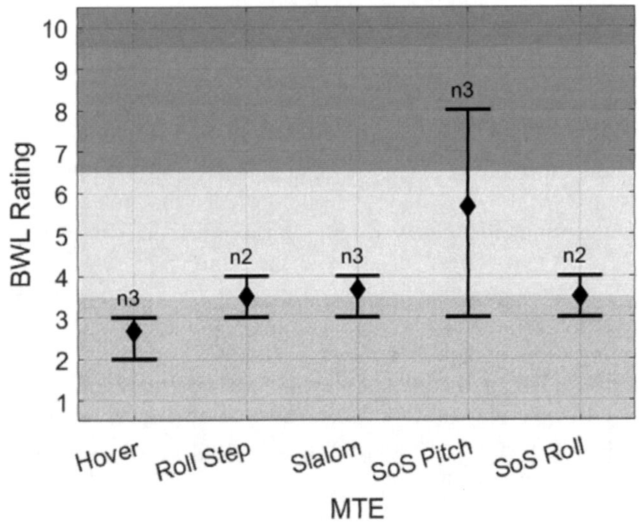

Figure 7: Summary of BWRs Obtained, Configuration: 4E

Figure 8 shows the Bedford Workload ratings awarded during the Hover MTE with respect to each inceptor force feel configuration. As shown, case 1D received BWRs within Level 2. All other configurations were generally awarded Level 1 HQs. For the Hover MTE, all pilots generally felt that configuration 1D was not suitable for the task. This was reflected in ratings awarded and through the quantitative performance. For example, regarding configuration 1D, Pilot A was not able to remain within the adequate tolerances (drifting outside) and had trouble correcting both lateral and longitudinal drift. All three pilots disagreed (or strongly disagreed) that the pitch breakout force was desirable for the given MTE. All three pilots also made comments that the stick (inceptor) felt "too heavy" or "sluggish." Some general comments from the pilots included disharmony, sluggish, heavy. These deficiencies led to a decrease in task performance.

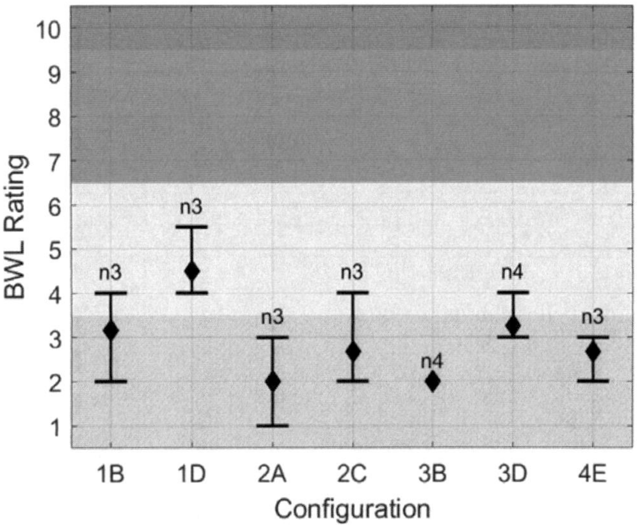

Figure 8: Bedford Workload Ratings, Hover

Configuration 3B, (critically damped, high natural frequency), was found to have some of the most favorable characteristics for the Hover MTE. Based on Pilot comments, the much lighter stick feel resulted in the expected improved performance over configuration 1D. Configuration 2A was also found to be generally favorable by all pilots, receiving a HQR 1 from one of the pilots.

For configuration 1B, large differences in pilot opinion were found. Pilot C reacted positively to this configuration awarding HQR 1. The other pilots awarded ratings in the Level 2 region. The pilot-to-pilot discrepancy is believed to be due to the strategy employed by Pilot C during the Hover. Due to the HQs of the vehicle with the given feel-system configurations, and lack of disturbances, Pilot C could achieve the desired performance with minimal control inputs. Therefore, despite this configuration being labelled 'heavy', Pilot C did not experience discomfort during completion of the task.

Scalograms allow for the visualization of pilot activity when using the control inceptor and provide rapid insight into the "peaks" in pilot activity. As depicted in Figure 9, scalograms show the power in the inceptor input signal with respect to time and frequency. Scalograms have been shown to be a useful tool in differentiating run-to-run and pilot-to-pilot characteristics regarding pilot-vehicle system behavior [13]. The time varying Scalogram technique is described by Thompson *et.al.* in Reference [14], and Mallet in Reference [15].

Scalogram analysis further demonstrates the difference in performance between two of the inceptor force feel configuration cases; 1D and 3B. Figure 9 and Figure 10 show the lateral inceptor input for the Hover MTE flown by Pilot A for cases 1D and 3B respectively. The only variable between the cases is the force-feel setting. As shown for 1D (Figure 9), strong peaks between 1-10 rad/s were observed throughout the MTE (approximately 60 seconds). These peaks increase in magnitude during the stabilization and hover element of the task (after 30 seconds). In contrast, case 3B (Figure 10), shows very low inceptor control activity and no peaks at higher frequency. This correlates with pilot feedback, that he was very light on the controls, applying small corrective inputs for the 3B configuration.

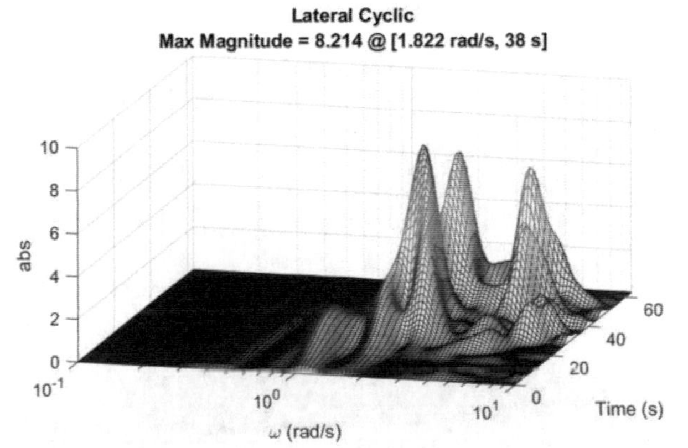

Figure 9: Scalogram, Hover, Control Inceptor Input, Lateral, Configuration 1D, HQR 5, BWL 5.5, PIOR 2

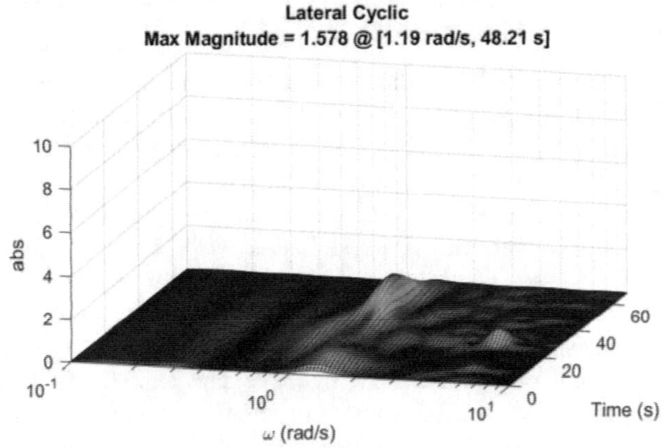

Figure 10: Scalogram, Hover, Control Inceptor Input, Lateral, Configuration 3B, HQR 2, BWL 2, PIOR 2

Roll Step MTE

The Roll Step MTE is a lateral maneuver that requires the pilot to initiate the maneuver in level unaccelerated flight at the centerline between two gates. The pilot must then perform a turn to traverse laterally 200 ft, stabilize, then perform the turn in the opposite direction. The maneuver is completed at a given height and flight speed. The maneuver is completed when the aircraft successfully travels through all gates. The flight speed can be used to modify the task aggression.

Figure 11 shows the HQRs awarded with respect to the configuration. Similar to the Hover MTE, the majority of HQRs awarded were Level 1. For the Roll Step MTE, configuration 3B was found to be the most favorable. For cases 1B, 1D, and 4E HQ ratings did not discern a difference between the cases. However, further feedback from the pilot questionnaire indicated that case 1D was the least suitable for the Roll Step MTE, as was the case for the 1D Hover MTE.

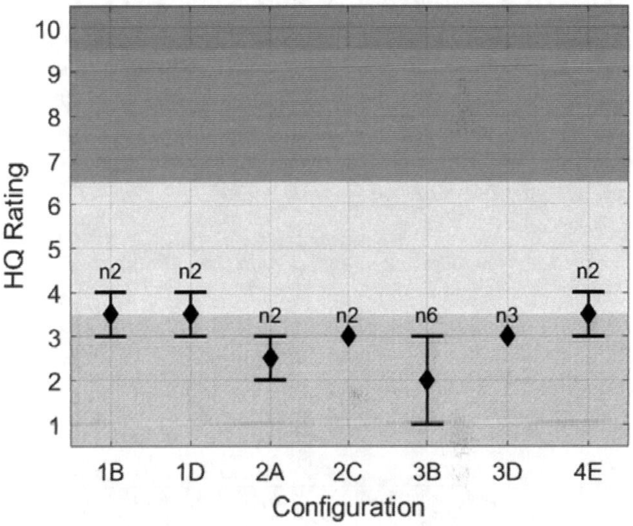

Figure 11: HQRs, Roll Step MTE

The Roll Step MTE successfully exposed differences between the inceptor force feel configurations. For the Roll Step MTE, the pilots preferred a lighter stick feel than configuration 1D. Pilot C stated that it took more movement to displace the stick in case 1D and rated the Roll Step task with a HQR 3 and BWL 4. Pilot A stated that the 1D configuration response was slow, "laggy", and heavy. The pilot also stated that he found it challenging to maintain the centerline. Pilot A also stated that in configuration 1D, the aircraft "flies like a bomber," awarding HQR 4 for the Roll Step task. For configuration 1D, the questionnaire results indicated that pitch and roll sensitivity were not desirable, that there was disharmony between the axes, and that the aircraft response lagged the control input.

In contrast, case 3B had favorable comments. The pilot stated that the harmony felt nice, predictable, and that the aircraft responded well. Pilot A stated that configuration 3B was their favorite, also mentioned multiple times that it felt "snappy." However, both pilots stated that the pitch axis felt heavy. Pilot

A also stated that case 3B had heavy stick breakout force but agreed that 3B was a generally desirable inceptor configuration for completing the Roll Step task. For the Roll Step task, HQR 2-3 and BWL 2-3 and PIOR 1 were assessed across the pilots.

For case 2A Pilot C stated the configuration was "more comfortable" than case 3B, stating the aircraft was "fast through the gates." Pilot A felt that there was free-play in the lateral axes but said that the system had a snappy response and was able to perform the task correctly, with level-wing through the target gates.

Figure 12 and Figure 13 show the lateral cyclic scalograms for configuration 1D and 3B, while Figure 14 shows the scalogram for the longitudinal cyclic of configuration 1D. Cases 1D and 3B represent the least favorable inceptor configurations for the roll axis. For all three pilots, the Roll Step task is dominated by the lateral cyclic response with two unique frequencies dominating the lateral response. As seen in Figure 12 and Figure 13, the lower frequency peak is likely associated with the Roll-Step's acquisition of an initial new lateral position, while the second slightly higher frequency peak appears to be associated with the precision capture of the new position. This was apparent for all pilots, regardless of feel system configuration.

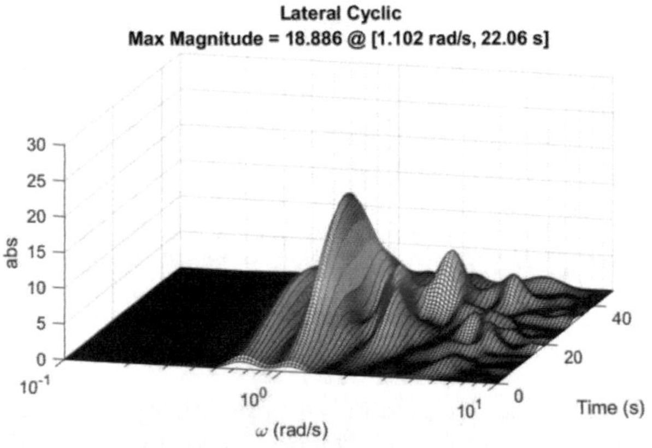

Figure 12: Lateral Cyclic, Roll Step, Configuration 1D, HQR 4, BWL 4, PIOR 2

The Slalom MTE is as maneuver that is initiated in level unaccelerated flight with the front of the aircraft lined up with the centerline of the test course. The pilot must perform a series of smooth turns at 500-ft intervals (at least twice to each side of the course). The turns are 50 ft from the centerline, with a maximum allowed lateral error of 50 ft. The maneuver is accomplished below a given reference altitude.

Referring to Figure 15, the Slalom MTE (like the previous Hover and Roll Step maneuver tasks), achieved HQRs in the Level 1 region, but with several cases awarded Level 2 ratings. All pilots generally agreed that configuration 1D was not suitable for the Slalom task. In particular, Pilot B strongly indicated that configuration 1D was unsuitable, awarding an HQR 5 for that case. In general, pilot feedback was positive for inceptor force-feel configuration 3B. However, during the initial 3B evaluation of the Slalom MTE, Pilot B felt there was insufficient stick travel to complete the MTE. When the pilot repeated the task, ratings were positive. For this reason, a large spread of HQRs was found for configuration 3B, as seen in Figure 15.

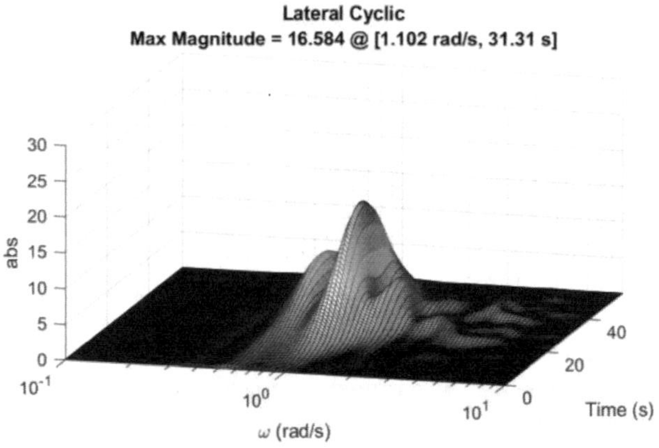

Figure 13: Lateral Cyclic, Roll Step, Configuration 3B, BWL 3, PIO 1

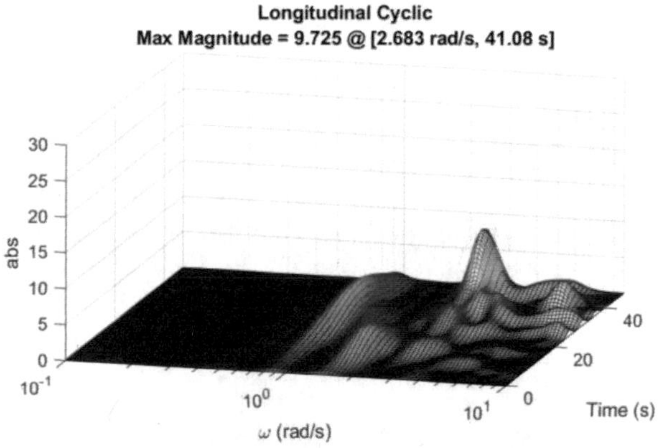

Figure 14: Longitudinal Cyclic, Roll Step, Configuration 1D, HQR 4, BWL 4, PIOR 2Slalom MTE

The Slalom MTE indicated the system had sufficient bandwidth to complete the required task. The performance plots (Figure 15) showed no major deficiencies for any of the inceptor configurations. Unlike the Hover and Roll Step MTEs, (see Table 2) two additional inceptor configurations (3A and 3C) were flown for the Slalom MTE. For the Slalom MTE, inceptor configuration 3C was found to be suitable, which may be of interest for future evaluations. Overall, during testing, the pilots preferred evaluating the Roll Step MTE compared to the Slalom MTE. The Roll Step MTE included both low-frequency translational inputs and higher frequency stabilization and tracking, which was not generally the case for the Slalom. For the Slalom MTE, only low frequency inputs could be used to achieve the desired performance.

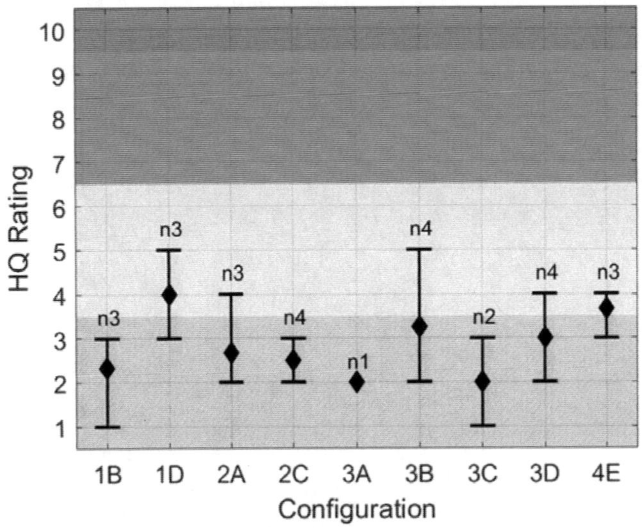

Figure 15: HQRs, Slalom MTE

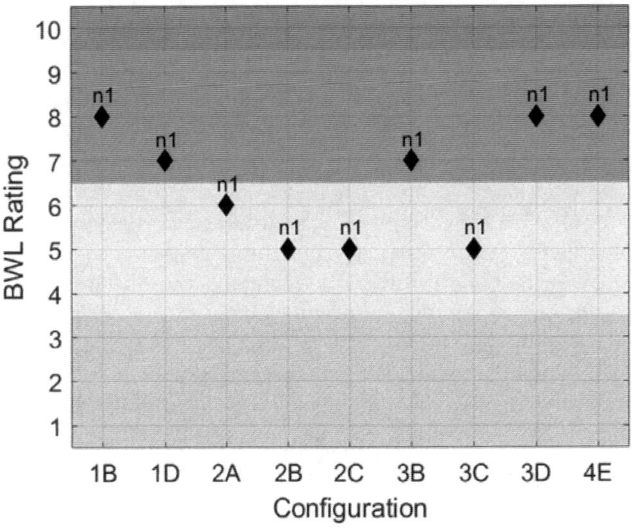

Figure 16: Bedford Workload Ratings Awarded for SoS Pitch Tracking, Pilot B.

Sum-of-Sines (SoS) Pitch Tracking MTE

The Sum-of-Sines Pitch MTE asks the pilot to aggressively track the displayed signal in pitch attitude while keeping the aircraft wings level and while attempting to keep tracking errors within the specified tolerances. This task is driven by an automated command signal selected by the flight test engineer.

The Sum-of-Sines Pitch tracking MTE was successful in exposing differences between the inceptor configurations. In several instances, inceptor differences, regarding MTE performance, was apparent. From all the MTEs attempted, the SoS was found to have the poorest ratings, both in terms of HQR and BWR awarded. With particular regard to the BWRs, several ratings were found at Level 3. Pilot B gave the poorest BWRs for the SoS MTE. The ratings awarded, with respect to inceptor force-feel configuration, are shown in Figure 16. The pilot stated that high concentration was required to complete the task and performance was difficult to achieve. Pilots generally achieved desired (or adequate) task performance, thus the high (poor) BWRs were not consistent with task performance. Evidently the high task workload was the main factor in the BWR score and not task performance requirements. For this reason, a larger spread was found in the BWRs compared to the HQRs for the same (SoS) MTE. Generally, HQRs for the SoS task were within the Level 2 region.

Based on the SoS MTE feedback from the pilot questionnaires, it appears that the perceived 'breakout force' was too large for the inceptor force-feel configurations, which may have been a factor that led to high workload with this configuration. The generally high workload was also due to high levels of concentration required to complete the task and maintain tracking performance for over 60 seconds. Some of the general comments from the pilots included "too much force, exhausting, and heavy". The pilots stated that they were working harder, which led to poorer tracking.

The SoS MTE tracking performance evaluation runs for two of the pilots, Pilot A and Pilot C, are shown in Figure 17-Figure 20. These evaluations represent inceptor configuration 1D (Figure 17 and Figure 18) and inceptor configuration 3D (Figure 19 and Figure 20).

For the SoS Pitch and Roll Tracking tasks, a 'score' (or 'performance') can be calculated based on the overall time that the pilot maintains adequate and/or desired performance. The adequate score is the percentage of time for which the pilot maintained the pitch (or roll) attitude within the adequate boundaries. For inceptor configuration 1D, Pilot A (Figure 17) was able to achieve a relatively high adequate performance of greater than 90%, while Pilot C (Figure 18) preformed much worse reaching only 67.24% adequate performance throughout the maneuver duration. As shown in Figure 18, large oscillations are present throughout. These appear to be pilot-induced oscillations (PIO), even though these were not identified by the pilot (PIOR 1). There are clear oscillations throughout and desired performance cannot be sustained. These oscillations were noted by the test team during the evaluations.

Figure 19 and Figure 20 show generally improved SoS Pitch tracking results for the inceptor force feel case 3B, where Pilot A and C both improved desired performance, and Pilot C also improved adequate performance. For this case, some of the pilot comments included "predictable" and "more responsive". Configuration 3B was one of the preferred cases for Pilot A, receiving the best (lowest) HQR. However, Pilot A generally felt the 'breakout forces' were too high, and that the inceptor control configuration generally did not have a light quick response. For Pilot C, oscillations appear to be present in the maneuver (as with the configuration 1D case) however these oscillations appear to be smaller for inceptor configuration 3B compared to the inceptor 1D case. For this reason, the inceptor 3B configuration's adequate performance is better than the configuration 1D performance.

For the SoS MTE, the questionnaire feedback (from all pilots) indicates that the perceived 'breakout force' for inceptor configurations 1D and 3D was too large and did not favor the high workload SoS task. The break-out force may have been a factor that led to high workload, alongside the generally high level of concentration required to complete the task and maintain tracking for over 60 seconds. Some of the general comments from the pilots included too much force, exhausting, and heavy. The pilots stated that they were working harder, which led to poorer tracking.

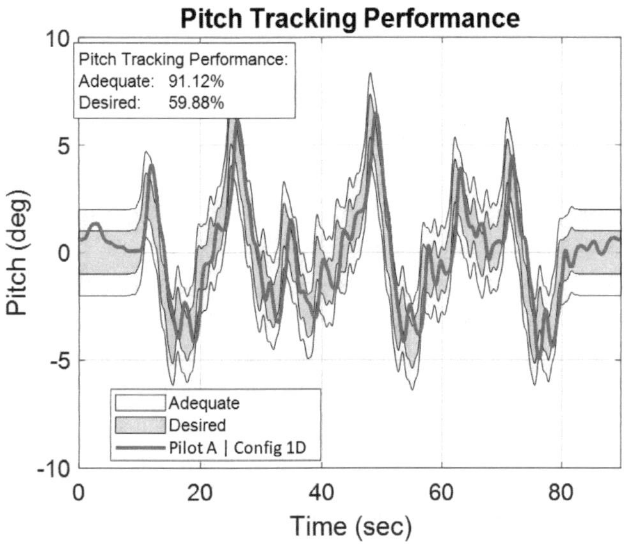

Figure 17: Example of Pitch Tracking Performance, Pilot A, Configuration 1D

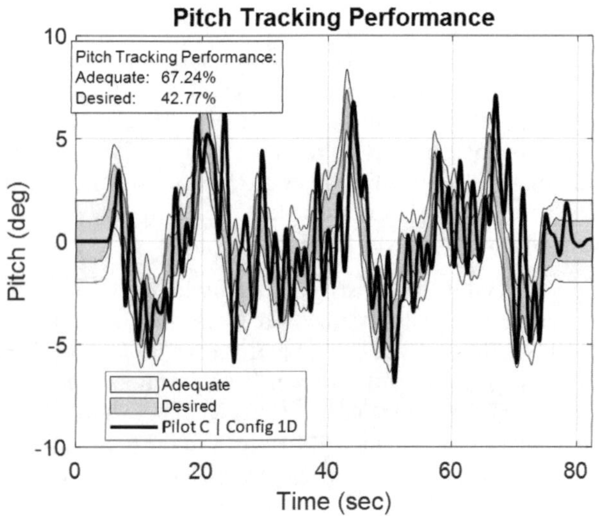

Figure 18: Example of Pitch Tracking Performance, Pilot C, Configuration 1D

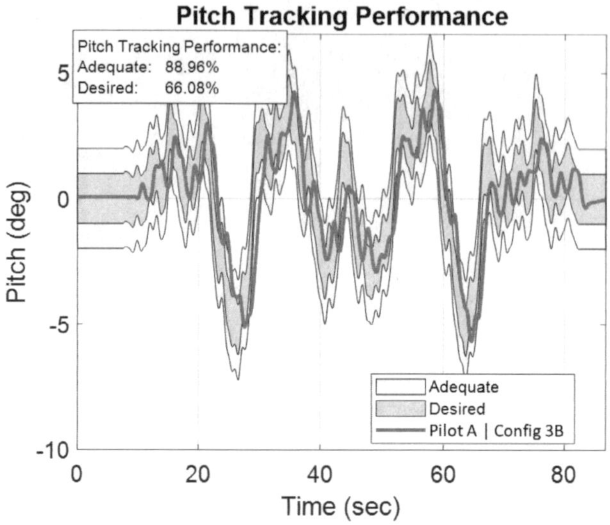

Figure 19: Example of Pitch Tracking Performance, Pilot A, Configuration 3B

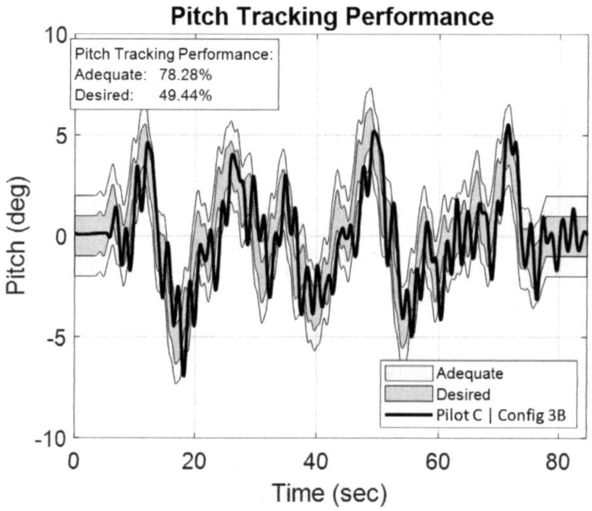

Figure 20: Example of Pitch Tracking Performance, Pilot C, Configuration 3B

Overall, pilots preferred the cases with lighter stick (inceptor) forces, a factor that was generally found to significantly reduce workload. This result is also supported with evidence that the inceptor oscillations observed in Pilot C tracking were reduced with the lighter forces.

Sum-of-Sines (SoS) Roll Tracking MTE

The Sum-of-Sines Roll MTE asks the pilot to aggressively track the displayed signal in roll attitude while keeping the aircraft at a constant pitch angle and while attempting to keep tracking errors within the specified tolerances. This task is driven by an automated command signal selected by the flight test engineer.

While still informative, the Sum-of-Sines Roll tracking MTE was less successful in exposing key differences between inceptor force feel configurations (compared to Sum-of-Sines Pitch tracking MTE). Figure 21 shows the SoS Roll MTE BWRs with respect to inceptor force feel configuration, for all three pilots. As shown, the best HQ was obtained for the case 3B, which also had the highest spread (and was flown the greatest number of times). Configuration 2A was also found to exhibit low workload, however some oscillations were apparent. The inceptor force-feel configuration 4E was found to exhibit harmony issues.

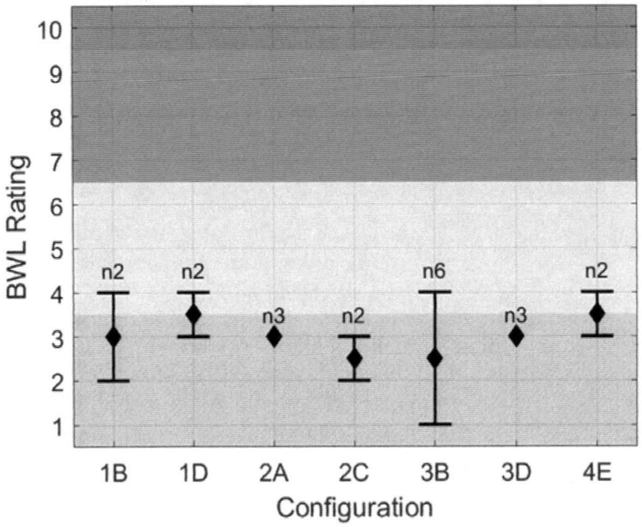

Figure 21: Overall BWRs obtained from the SoS Roll Tracking MTE

DISCUSSION

Based on the overall results, the Hover and Sum-of-Sines MTEs were the most successful tasks for highlighting the impact of the different inceptor force-feel configurations. Pilot-induced oscillations were observed in the Sum-of-Sines task for both pitch and roll axes, while the Hover task highlighted force-feel configurations where control system lag was more noticeable by the pilots. Lag or noticeable delay in the Hover task resulted in increased drift, requiring more compensation from pilots (more frequent/aggressive pilot corrections). The Slalom and Roll Step were useful for comparing inceptor force-feel configurations and offered pilot comments on the agility of the aircraft.

Bandwidth and phase delay [9] represent additional measures that can be used to characterize the speed of aircraft response with respect to pilot stick input. Bandwidth is usually calculated on the basis of the inceptor position but can be computed using the inceptor force signal to shows the effect of the different inceptor force-feel settings. Phase delay and bandwidth results, based on inceptor force, are shown in Figure 22 (for the SoS Pitch MTE) and Figure 23 (for the SoS roll MTE). Inceptor force-feel configurations 2A and 3B are located within the Level 2 region, near the Level 1 boundary (for both pitch and roll axes). These inceptor configurations are associated with a less sluggish response and reduced latency compared to the other tested inceptor force-feel configurations. In contrast, configurations 1D, 4E, and 1B are located within the Level 3 region or near its boundary and are associated with a more sluggish response and higher latency. This result is in-line with the performance seen during the SoS MTE tasks and the pilot comments obtained during debriefing. Overall, the phase delay/bandwidth results illustrate a trend, which is that inceptor force-feel configurations with a lower damping ratio and higher natural frequency tend to bring the system closer to Level 1 HQ ratings. In contrast, configurations with a higher damping ration and lower natural frequency result in a more sluggish response and higher latency.

Considering the aircraft state time histories, Pilot C frequently experienced pilot-induced oscillations. These were most noticeable during the Sum-of-Sines maneuvers but are also evident some of the other MTEs. Nevertheless, pilot C did not typically indicate the presence of the PIOs in his ratings. It is not uncommon for PIO to go unrecognized by pilots when flying in a fixed-base simulator. In some cases, such as SoS Pitch (inceptor force-feel configuration 1D) the, PIO was quite severe given the tracking performance, but the pilot rated it with a PIOR 1. The pilot did give the 1D inceptor configuration a higher BWLR and HWR indicating the pilot was aware of the extra difficultly in completing the task but did not attribute this to PIO.

Pilot B often gave consistently higher ratings for all the configurations compared to pilots A and C. This was especially true during the scoring of the sum-of-sines pitch tracking MTE. This may have been due to the fact that pilot B is the "lower gain" or "lower bandwidth" pilot, when compared to the other two, hence the tasks that demand more continuous compensation result in higher ratings for Pilot B.

Overall, the most favorable ratings were awarded for configuration 3B. This configuration corresponds to an inceptor with high natural frequency and low damping ratio. The favored configuration (3B) has a natural frequency of 20 rad/s and a damping ratio of 1. The test results indicate that a slightly lower natural frequency (i.e., 15 rad/s) was also acceptable in most tasks. High damping increases the pilot's response lag and results in a correspondingly worse rating for the associated HQR, BWL, and PIO. The three pilots generally agreed that configuration 1D felt too heavy across all tasks, and thus received the worst average Handling Qualities and Bedford Workload ratings. Based on the overall results, configurations with a frequency between 15-20 rad/s and a damping ratio between 1-2 were found to be acceptable, for the rate command response. This corresponds to a Level 2 (bordering on Level 3 for the favored configuration 3B) in the originally proposed boundaries seen in Figure 2.

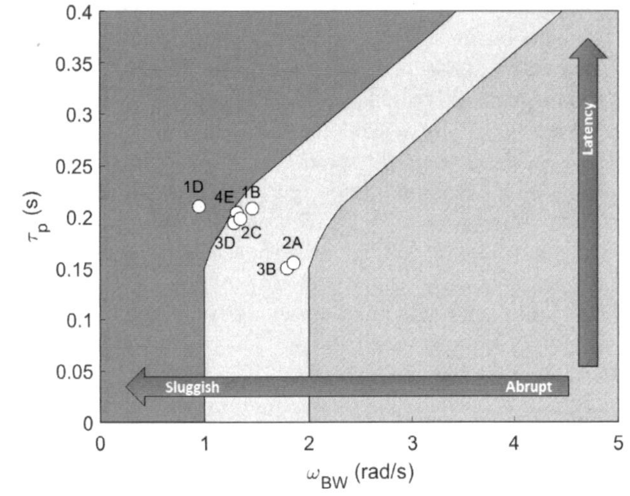

Figure 22: Pitch Axis Bandwidth/Phase Delay

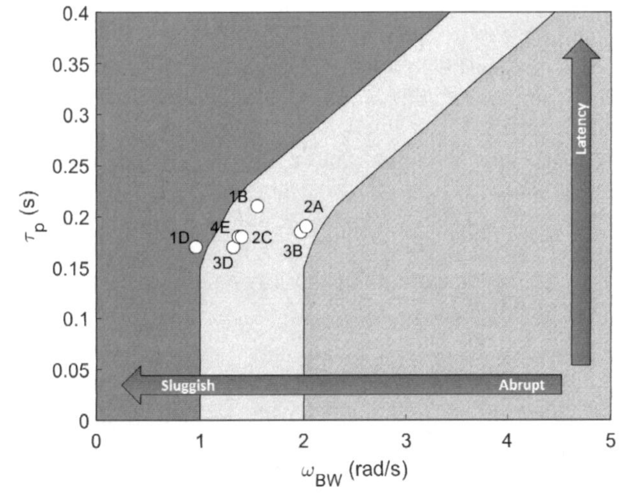

Figure 23: Roll Axis Bandwidth/Phase Delay

Configuration 3D was a test point that fell inside of the originally proposed Level 1 region. Configuration 1D consistently lead to poorer aircraft tracking performance, when compared to the so called "Level 2" configurations. Additionally, 1D was the only configuration that was Level 3 for both roll and pitch axis in bandwidth/phase delay. Based on the evaluations and analyses performed in this study, it is recommended the boundaries be revisited in further studies to investigate the applicability for different control response types and for sidestick inceptors.

CONCLUSIONS

The test campaign described in this paper was designed to investigate the influence of inceptor force-feel characteristics, for one flight control response type, tested in a number of mission task elements (MTEs). Key conclusions from this study are as follows:

- Inceptor force-feel configurations with lower damping and higher natural frequency resulted in lower latency and a less "sluggish" aircraft response. Overall maneuver task performance was improved for inceptor force-feel configurations with a natural frequency range of 15-20 rad/s, and damping ratios of 1-2.
- The multi-axis nature and the three distinct elements of the Hover MTE allowed for a rich exploration of the pilot inputs as inceptor force-feel characteristics were varied. Inceptor trends could be observed with consistency among the pilots.

For the Slalom MTE, the lateral cyclic scalograms for all three pilots were similar in frequency and amplitude as the nature of the task dictates the required pilot inputs.

Generally, results from this study reflect those from previous campaigns and results available in the literature. A trend was found towards improved Handling Qualities with increasing natural frequency and damping of the force-feel characteristics. The natural damping due to the pilot's arm position for the sidestick inceptors appeared to reduce the requirements for damping in the force-feel system. Results obtained strongly reflected those from DLR studies, where points with high natural frequency and low damping were found to be acceptable. It is important to note that all tests reported here were with a fixed-motion base simulator. The preferred configurations with respect to damping ratio are expected to increase once motion is introduced into the flight simulator. Thus, the recommendation is to consider damping ratios not only at the preferred level (found in this study with case 3B), but also slightly higher values, that is, damping ratios between 1 and 2.

Author Contact: Michael Jones mjones@systemstech.com

ACKNOWLEDGMENTS

The authors would like to thank all who supported the successful completion of the piloted simulation campaigns. In particular, the authors would like to thank the three test pilots who participated for the feedback and engagement during the trials. In addition, the authors would like to thank Dr. Jeffrey Lusardi, for technical support and guidance during the investigations.

APPENDIX

Table 4: Pilot Debrief Questionnaire

Question	Strongly Disagree	Disagree	Neither Agree nor Disagree	Agree	Strongly Agree
Q1. The cyclic control pitch/roll harmony was desirable for the selected MTE.	O	O	O	O	O
The cyclic characteristics identified below were desirable for the selected MTE:					
• **Q2.** Pitch sensitivity (gradient)	O	O	O	O	O
• **Q3.** Roll sensitivity (gradient)	O	O	O	O	O
• **Q4.** Pitch breakout force	O	O	O	O	O
• **Q5.** Roll breakout force	O	O	O	O	O
Q6. The damping of the cyclic response in pitch was ideal for the selected MTE.	O	O	O	O	O
Q7. The damping of the cyclic response in roll was ideal for the selected MTE.	O	O	O	O	O
Q8. The cyclic inceptor in pitch had a light, quick response	O	O	O	O	O
Q9. The cyclic inceptor in roll had a light, quick response.	O	O	O	O	O
Q10. There was little or no perceived delay in the aircraft response to control input.	O	O	O	O	O
Q11. The stick travel in the pitch and roll axes was sufficient to perform the selected MTE	O	O	O	O	O

Table 5: Hover MTE Pilot Ratings

MTE	Config	Pilot	HQR	BWL	PIOR	Q1	Q2	Q3	Q4	Q5	Q6	Q7	Q8	Q9	Q10	Q11
Hover	1B	A	4	3.5	2	0	1	1	1	1	1	-1	-1	-1	1	1
Hover	1B	B	5	4	3	1	-1	-1	0	0	0	0	0	0	1	2
Hover	1B	C	1	2	2	1	1	1	1	1	1	1	1	1	1	1
Hover	1D	A	5	5.5	2	-1	1	1	-1	-1	-1	-1	-1	-1	1	1
Hover	1D	B	5	4	2	-1	0	0	-2	-2	1	1	-2	-2	1	1
Hover	1D	C	5	4	3	1	-1	-1	-1	-1	0	0	0	0	-1	1
Hover	2A	A	2	2	2	1	1	1	1	1	1	1	0	0	0	1
Hover	2A	B	3	3	2	1	1	1	1	1	1	1	1	1	0	1
Hover	2A	C	1	1	1	1	1	1	1	1	1	1	1	1	1	1
Hover	2C	A	3	2	2	1	1	0	1	1	1	1	1	1	0	1
Hover	2C	B	5	4	2	0	0	1	1	1	1	0	1	1	0	1
Hover	2C	C	2	2	1	1	1	1	1	1	0	0	1	1	1	1
Hover	3B	A	2	2	2	1	1	1	1	1	1	1	1	1	1	1
Hover	3B	A	2	2	1	1	1	1	1	1	1	1	1	1	1	1
Hover	3B	B	2	2	2	2	2	2	2	1	2	2	2	2	1	2

MTE	Config	Pilot	HQR	BWL	PIOR	Q1	Q2	Q3	Q4	Q5	Q6	Q7	Q8	Q9	Q10	Q11
Hover	3B	C	3	2	2	2	1	1	1	1	0	0	1	1	1	2
Hover	3D	A	5	4	2	0	1	1	0	1	-1	-1	-1	1	-1	1
Hover	3D	B	3	3	2	1	1	0	1	1	1	0	0	0	0	2
Hover	3D	B	3	3	2	2	1	1	-1	-1	1	1	-1	-1	1	1
Hover	3D	C	4	3	2	1	0	0	0	0	0	0	0	-1	-1	1
Hover	4E	A	3	3	2	1	1	1	1	0	0	1	-1	1	1	1
Hover	4E	B	3	3	2	1	1	1	1	1	1	0	0	0	2	2
Hover	4E	C	3	2	2	1	2	0	1	0	1	1	1	1	1	2

Table 6: Roll Step MTE Pilot Ratings

MTE	Config	Pilot	HQR	BWL	PIOR	Q1	Q2	Q3	Q4	Q5	Q6	Q7	Q8	Q9	Q10	Q11
Roll Step	1B	A	4	3.5	2	-1	-1	-1	-1	1	-1	0	1	0	0	1
Roll Step	1D	A	4	4	2	-1	-1	-1	0	0	-1	0	-1	-1	-1	1
Roll Step	1D	C	3	4	1	1	1	1	1	0	0	0	0	0	0	1
Roll Step	2A	A	3	3	2	1	1	1	1	1	1	1	-1	1	1	1
Roll Step	2A	C	2	3	1	1	1	1	0	0	1	1	0	0	0	1
Roll Step	2C	A	3	3	2	-1	-1	1	-1	0	-1	-1	-1	1	0	1
Roll Step	2C	C	3	3	1	1	1	1	0	1	0	1	0	0	0	1
Roll Step	3B	A	3	3	1	1	1	1	1	1	1	1	1	1	1	0
Roll Step	3B	A	2	2	1	1	1	1	1	1	1	1	0	0	1	1
Roll Step	3B	C	2	3	1	1	1	1	1	1	0	0	0	1	0	1
Roll Step	3B	C	2	2	1	1.5	1.5	1.5	1.5	1	1	1	1	1	1	1
Roll Step	3B	C	2	3	1	1	1	1.5	1.5	1	1	1	1	1	1	1
Roll Step	3D	A	3	3	1	0	1	1	1	1	0	0	1	1	1	1
Roll Step	3D	A	3	3	1	-1	-1	0	-1	-1	-1	1	1	0	1	1
Roll Step	3D	C	3	3	1	1.5	1	1.5	1	1	1	1	1	1	1	1
Roll Step	4E	A	3	3	1	1	1	0	1	1	1	0	0	0	0	1
Roll Step	4E	C	4	4	1	1	0	0	0	0	0	0	1	1	0	1
Roll Step	1B	C	3	4	1	1	1	1	1	1	1	1	1	1	1	1
Roll Step	3B	C	1	2	1	1.5	1.5	1.5	1.5	1.5	1	1.5	1	1	1.5	1

Table 7: Slalom MTE Pilot Ratings

MTE	Config	Pilot	HQR	BWL	PIOR	Q1	Q2	Q3	Q4	Q5	Q6	Q7	Q8	Q9	Q10	Q11
Slalom	1B	A	3	3	1	0	-1	1	1	1	-1	1	1	1	1	1
Slalom	1B	B	3	3	1	1	1	1	1	1	1	1	1	1	1	0
Slalom	1B	C	1	1	1	2	2	2	2	2	2	2	1	1	1	2
Slalom	1D	A	4	3	1	-1	-1	1	0	0	-1	1	-1	0	-1	1
Slalom	1D	B	5	6	1	0	-2	-2	-2	-2	-1	-1	-1	-2	-1	-2
Slalom	1D	C	3	4	1	0	0	0	0	0	0	0	-1	-1	-1	1
Slalom	2A	A	2	2	1	1	1	1	1	1	1	1	0	0	1	1
Slalom	2A	B	4	4	1	0	-1	-1	1	2	0	-1	1	1	0	-1
Slalom	2A	C	2	2	1	1	1	1	1	1	1	1	1	1	1	1
Slalom	2C	A	2	2	-	1	1	1	1	1	0	1	1	1	1	1
Slalom	2C	B	3	4	1	0	1	1	1	1	1	1	1	1	0	-1
Slalom	2C	C	3	4	1	1	0	0	0	0	0	0	1	1	0	1
Slalom	2C	C	2	2	1	1	1	1	1	1	1	1	1	1	1	1
Slalom	3A	C	2	2	1	1	1.5	1.5	1.5	1.5	1	1	2	2	2	1

MTE	Config	Pilot	HQR	BWL	PIOR	Q1	Q2	Q3	Q4	Q5	Q6	Q7	Q8	Q9	Q10	Q11
Slalom	3B	A	3	2.5	1	0	1	1	1	1	0	0	-1	1	1	-1
Slalom	3B	B	5	6	1	-1	0	-1	0	0	-2	-2	-2	-2	-2	-2
Slalom	3B	B	3	2	1	1	1	2	1	1	1	2	1	1	1	0
Slalom	3B	C	2	2	1	1	1	1	1	1	1	1	1	1	1	1
Slalom	3C	A	3	3	1	0	0	0	1	1	-1	1	0	1	1	-1
Slalom	3C	C	1	1	1	2	2	2	2	2	1.5	1.5	1	1	1	1.5
Slalom	3D	A	2	2	1	-1	1	1	1	1	-1	1	-1	1	1	1
Slalom	3D	B	4	4	1	0	0	0	-1	-1	0	-1	-1	-1	-1	-2
Slalom	3D	C	3	3	1	1	1	1	1	1	1	1	0	0	0	1
Slalom	3D	C	3	2	1	1	0	0	0	0	0	0	-1	-1	0	1
Slalom	4E	A	3	3	1	1	1	-1	1	1	1	-1	0	0	1	1
Slalom	4E	B	4	4	1	0	-2	1	0	0	1	1	-1	-1	0	1
Slalom	4E	C	4	4	1	1	0	0	0	0	1	1	0	-1	0	1

Table 8: SoS Pitch MTE Pilot Ratings

MTE	Config	Pilot	HQR	BWL	PIOR	Q1	Q2	Q3	Q4	Q5	Q6	Q7	Q8	Q9	Q10	Q11
SoS Pitch	1B	A	3	2	2	1	1	-	1	-	-1	-	1	-	1	1
SoS Pitch	1B	B	5	8	2	0	-1	-	0	-	-1	-	-1	-	-1	0
SoS Pitch	1B	C	4	4	1	1	1	-	0	-	0	-	1	-	1	1
SoS Pitch	1D	A	3	3	2	0	-1	-	-1	-	1	-	-1	-	0	1
SoS Pitch	1D	B	4	7	1	0	0	-	-2	-	1	-	-2	-	0	0
SoS Pitch	1D	C	6	6	1	-	-1	-	-2	-	-1	-	-1	-	-1	0
SoS Pitch	2A	A	4	4	2	1	1	-	1	-	-1	-	1	-	1	1
SoS Pitch	2A	B	4	6	2	0	-1	-	2	-	1	-	0	-	0	0
SoS Pitch	2A	C	4	4	1	0	1	-	0	-	0	-	1	-	1	1
SoS Pitch	2B	B	4	5	1	0	1	-	1	-	1	-	1	-	1	0
SoS Pitch	2B	C	4	4	1	1	1	-	1	-	1	-	1	-	0	1
SoS Pitch	2C	A	4	4	2	0	1	-	1	-	-1	-	1	-	1	1
SoS Pitch	2C	B	4	5	1	0	0	-	-1	-	1	-	-1	-	0	0
SoS Pitch	2C	C	4	4	1	1	1	-	-1	-	0	-	1	-	0	1
SoS Pitch	3B	A	2	2	2	1	1	-	1	-	-1	-	1	-	1	1
SoS Pitch	3B	A	3	3	2	1	0	-	1	-	-1	-	0	-	0	1
SoS Pitch	3B	B	4	7	2	0	0	-	-1	-	-1	-	0	-	-1	0
SoS Pitch	3B	C	5	5	1	0	-1	-	0	-	-1	-	0	-	0	0
SoS Pitch	3C	A	4	4	2	1	0	-	0	-	-1	-	-1	-	-1	1
SoS Pitch	3C	B	4	5	1	0	1	-	1	-	1	-	1	-	1	0
SoS Pitch	3C	C	5	5	1	-	0	-	0	-	0	-	0	-	0	1
SoS Pitch	3D	A	2	2	2	1	1	-	1	-	0	-	0	-	1	1
SoS Pitch	3D	A	2	2	2	1	1	-	1	-	1	-	0	-	1	1
SoS Pitch	3D	B	5	8	1	0	-1	-	-2	-	0	-	-2	-	-1	0
SoS Pitch	3D	C	5	5	1	0	0	-	-1	-	-1	-	0	-	0	0
SoS Pitch	4E	A	3	3	2	1	1	-	1	-	0	-	0	-	1	1
SoS Pitch	4E	B	4	8	2	0	-1	-	-1	-	-1	-	-1	-	-1	1
SoS Pitch	4E	C	5	6	1	0	0	-	-1	-	-1	-	-1	-	-1	0

Table 9: SoS Roll MTE Pilot Ratings

MTE	Config	Pilot	HQR	BWL	PIOR	Q1	Q2	Q3	Q4	Q5	Q6	Q7	Q8	Q9	Q10	Q11
Roll Step	1B	A	4	3.5	2	-1	-1	-1	-1	1	-1	0	1	0	0	1

MTE	Config	Pilot	HQR	BWL	PIOR	Q1	Q2	Q3	Q4	Q5	Q6	Q7	Q8	Q9	Q10	Q11
Roll Step	1B	C	3	4	1	1	1	1	1	1	1	1	1	1	1	1
Roll Step	1D	A	4	4	2	-1	-1	-1	0	0	-1	0	-1	-1	-1	1
Roll Step	1D	C	3	4	1	1	1	1	1	0	0	0	0	0	0	1
Roll Step	2A	A	3	3	2	1	1	1	1	1	1	1	-1	1	1	1
Roll Step	2A	C	2	3	1	1	1	1	0	0	1	1	0	0	0	1
Roll Step	2C	A	3	3	2	-1	-1	1	-1	0	-1	-1	-1	1	0	1
Roll Step	2C	C	3	3	1	1	1	1	0	1	0	1	0	0	0	1
Roll Step	3B	A	3	3	1	1	1	1	1	1	1	1	1	1	1	0
Roll Step	3B	A	2	2	1	1	1	1	1	1	1	1	0	0	1	1
Roll Step	3B	C	2	3	1	1	1	1	1	1	0	0	0	1	0	1
Roll Step	3B	C	2	2	1	1.5	1.5	1.5	1.5	1	1	1	1	1	1	1
Roll Step	3B	C	1	2	1	1.5	1.5	1.5	1.5	1.5	1	1.5	1	1	1.5	1
Roll Step	3B	C	2	3	1	1	1	1.5	1.5	1	1	1	1	1	1	1
Roll Step	3D	A	3	3	1	0	1	1	1	1	0	0	1	1	1	1
Roll Step	3D	A	3	3	1	-1	-1	0	-1	-1	-1	1	1	0	1	1
Roll Step	3D	C	3	3	1	1.5	1	1.5	1	1	1	1	1	1	1	1
Roll Step	4E	A	3	3	1	1	1	0	1	1	1	0	0	0	0	1
Roll Step	4E	C	4	4	1	1	0	0	0	0	0	0	1	1	0	1

REFERENCES

[1] J. Lusardi, C. Blanken, C. Ott, C. Malpica and W. Gruenhagen, "In Flight Evaluation of Active Inceptor Force-Feel Charactersitics and Handling Qualities," in *American Helicopter Society 68th Annual Forum*, Forth Worth, TX, 2012.

[2] J. Hegg, M. Smith, L. Yount and J. Todd, "Features of active sidestick controllers," *IEEE Aerospace and Electronic Systems,* vol. 10, no. 7, pp. 31-34, 1995.

[3] M. Whalley, W. Hindson and G. Thiers, "A Comparison of Active Sidestick and Conventional Inceptors," in *American Helicopter Society 56th Annual Forum*, Virgina Beach, VA, 2000.

[4] T. Berger, M. Tischler and J. Horn, "High-Speed Rotorcraft Pitch Axis Response Type Investigation," in *Vertical Flight Society 77th Annual Forum*, Virtual, 2021.

[5] T. Schoenenberg, "Development of Rotorcraft Handling Qualities Criteria for Active Sidestick Force-Displacement Characteristics," in *American Helicopter Society 69th Annual Forum*, Phoenix, AZ, 2013.

[6] C. Blanken, M. Tischler, J. Lusardi, T. Berger, C. Ivler and R. Lehmann, "Proposed Revisions to Aeronautical Design Standard - 33E (ADS-33E-PRF) Towards ADS-33F-PRF," U.S. Army, Redstone Arsenal, AL, 2019.

[7] G. E. Cooper and J. R. P. Harper, "The Use of Pilot Rating in the Evaluation of Aircraft Handling Qualities," NASA TN D-5153, 1969.

[8] A. H. Roscoe and G. A. Ellis, "A Subjective Rating Scale for Assessing Workload in Flight. A Decade of Use," Royal Aerospace Establishment Technical Report TR90019, Farnborough, U.K., 1990.

[9] Anon., "Aeronautical Design standard Performance Specification Handling Qualities Requirements for Military Rotorcraft," US Army Aviation and Missile Command, Redstone, AL, 2000.

[10] D. A. DiFranco, "Flight Investigation of Longitudinal Short Period Frequency Requirements and PIO Tendencies," AFFDL-TR-66-163, 1967.

[11] M. Meyer and G. Padfield, "First Steps in the Development of Handling Qualities Criteria for a Civil Tiltrotor," in *American Helicopter Society 58th Annual Forum*, Montreal, Canada, 2002.

[12] D. H. Klyde, S. P. Pitoniak, P. C. Schulze, P. Ruckel, J. Rigsby, C. E. Fegely, H. Xin, W. C. Fell, R. Brewer, F. Conway, R. Mulato, J. Horn, L. C. R. Ott and C. L. Blanken, "Piloted Simulation Evaluation of Tracking Mission Task Elements for the Assessment of High-Speed Handling Qualities," *Journal of the American Helicopter Society,* vol. 65, no. 3, pp. 1-23, 2020.

[13] A. Lampton and D. H. Klyde, "Power Frequency – A Metric for Analyzing Pilot-in-the-Loop Flying Tasks," *Journal of Guidance, Control, and Dynamics,* vol. 35, no. 5, pp. 1536-1537, 2012.

[14] P. M. Thompson, D. H. Klyde, E. N. Bachelder and T. J. Rosenthal, "On-line Loss of Control Detection using Wavelets," NASA CR-2005-212873, 2005.

[15] S. Mallet, "A Wavelet Tour of Signal Processing," in *Academic Press*, San Diego, 1998.

Inter- and Intra- Pilot Rating Variability during Dynamic Interface Simulations

Jeremy Scroggs
Flight Dynamics Engineer
NAVAIR
Patuxent River, MD

Jonathan Dronfield
Flight Dynamics Engineer
NAVAIR
Patuxent River, MD

Zachary F. Smith
Flight Dynamics Engineer
NAVAIR
Patuxent River, MD

ABSTRACT

A large number of pilot-in-the-loop simulated landings to various ships were conducted with a rotorcraft model. Many of the runs exhibited small variations in wind azimuth and velocity, while several runs repeated previous wind conditions. For each landing, pilot workload values were recorded in terms of both Cooper-Harper Handling Quality Ratings (HQRs) and Deck Interface Pilot Effort Scale (DIPES). This study compares these ratings to highlight how repeatable a pilot's perceived workload is for the given task. The results are also compared against actual Dynamic Interface (DI) results. Overall, it is shown that the majority of the standard deviation values for these ratings range from 0.2 to 0.8. This is fairly high when the workload scales range from 1 to 10 and 1 to 5, for HQR and DIPES respectively, and even high when one considers that pilots very rarely supply DIPES ratings of 5 or HQRs of 1, 7, 8, 9, or 10. This warrants the suggestion that workload studies and Dynamic Interface trials should ensure data points are repeated an appropriate number of times to ensure confidence in the results.

INTRODUCTION

Historically, aircraft development has solely relied on live simulation, which is a real pilot flying a real aircraft in a real environment, but "simulating" the desired mission. However, recently, aircraft development has heavily leveraged the use of virtual and constructive simulation to decrease cost, increase safety, and improve the fielded solutions. These simulation techniques involve a real pilot flying a simulated aircraft in a simulated environment (i.e. pilot-in-the-loop simulation), and a simulated pilot (i.e. pilot model or pilot controller) flying a simulated aircraft in a simulated world, respectively. There is currently pressure to apply virtual and constructive simulation in establishment of aircraft envelopes to gain similar benefits. Launch and recovery envelopes (LREs) for aircraft desiring clearance to operate to United States of America (US) Navy ships have been established almost entirely through live flight test, entailing large expenses and significant logistical efforts to organize the test. Resources are often rendered unavailable to the US Navy fleet for extended periods, and test teams often have to wait for representative weather conditions to obtain maximum capability (Ref. 1). The US Navy has been developing methodologies that combine shore-based high-fidelity flight simulation with live testing, eventually allowing for a substantial reduction in live test operations as well as exploration of weather conditions not available during live test (Ref. 2). This new methodology is intended to provide the US Naval fleet with more operational capability. It could also have significant cost and agility benefits by significantly reducing or, in some cases, eliminating expensive and time-consuming live Dynamic Interface (DI) testing.

In order to apply this new methodology, it is important for the data stakeholders to understand the applicability and limitations of the data, such as qualitative pilot workload, especially when in the shipboard environment. The implications of how consistent pilot workload (actual and perceived) is at a given condition in the simulator environment, as well as a pilot's capability to discern differences between one condition and another, are crucial to understanding the required test point density of a simulation trial. Typically, this workload is determined through some form of qualitative pilot rating. It is well documented that there is a certain amount of non-repeatability in pilot workload ratings. Wilson and Riley outlined such variability for the Cooper Harper Handling Qualities Rating (HQR) 30 years ago (Ref. 3 and 4). However, it is not well known how repeatable and reliable a pilot's perception of workload and performance are while completing a landing to a ship, which historically uses a different rating scale. McDonnell cautions that any workload scale will have repeatability issues (Ref. 5). Still, there is hope that the more coarse Deck Interface Pilot Effort Scale (DIPES), a common DI workload rating scale used by the US Navy and US Marine Corp and several other nations (Ref. 1), will tighten any distribution. To use these simulations for their intended purpose of augmenting or replacing live DI testing, it is essential to be able to scope out the amount of required data for any piloted simulator event. More specifically, it needed to be established if, when performing LRE development similar to the one outlined in Ref. 6, it is prudent to involve multiple pilots or repeat a given test condition multiple times.

Presented at the VFS Handling Qualities Technical Meeting, Huntsville, Alabama, USA Feb. 7-8 2023.

Objectives of Present Work

To explore this issue, a Naval Innovation Science and Engineering (NISE) project, called Pilot Workload Sensitivity Study, was conducted during the 2021 and 2022 fiscal years. The US Navy rotorcraft was chosen as the test aircraft for the simulator events due to the availability of rich shipboard flight test data and the high fidelity simulation environment. This simulation environment was located at the Manned Flight Simulator (MFS) facility at Naval Air Station Patuxent River.

The objective of this project was to study the ability of pilots to consistently rate the ease (or difficulty) of fleet operations aboard ships by performing dynamic interface test events in a simulator environment. For each recovery, the team collected a series of pilot workload ratings. The analysis of these results initially intended to explore both the level of repeatability and sensitivity of these ratings. In this context, repeatability was defined as the likelihood of giving the same or similar rating at the same conditions, and sensitivity was defined as the ability to detect changes in conditions.

METHODOLOGY

MFS' facility consisted of a simulator cab, which was functionally identical to the real rotorcraft cockpit, and a control room with the required support equipment (Ref. 7). The flight dynamics model used for this event was a government owned, physics-based model that has been updated continuously for decades. This model is approved for and implemented in US Navy training devices and is the most accurate flight dynamics model for this rotorcraft in the world. The simulation featured a blade element model and incorporated Computational Fluid Dynamics (CFD)-based ship airwakes to replicate the shipboard environment (Ref. 8). The aircraft's flight control computers were emulated for this event, though the control laws were identical. Approaches to three unique US Navy ships were conducted, these ships will be hereby referred to as: Ship 1, Ship 2 and Ship 3. Three pilots provided formal ratings for this project. All three are US Naval Test Pilot School trained pilots, currently active members of HX-21, the US Navy's rotorcraft test squadron, and assigned to the tested rotorcraft as their primary platform. A US Navy or US Marine crew chief was also present for each flight event. The crew chief wore a virtual reality headset that allowed them to communicate real-time positioning information to the pilot, similar to how they would in the real aircraft. This assisted the pilot with cueing and positioning information.

Each pilot flew a short approach to the designated spot, starting approximately 1500 feet from the spot, and performed a nominal hover for about 20 seconds before landing. No requirement for hover time was specified; pilots were asked to complete the recovery as they would during a live DI event. Pilots were given up to 4 practice runs at the beginning of each flight period that were not scored. After these practices, all runs were scored. For each run, the pilot was asked to give four workload ratings: a Hover DIPES rating, a Landing DIPES rating, a Hover HQR, and a Landing HQR. The DIPES rating scale was collaboratively developed by The Technical Cooperation Program nations under Technical Panel 5 in order to accurately document pilot workload during launch and recovery and determine if that workload was acceptable for a fleet operator to perform the same task. DIPES ratings are one of the primary data sets that US Navy Flight Dynamics Engineers use in order to clear LREs for fleet operations. More on the origins of the DIPES rating scale and its use can be found in Ref. 1 and 9. The Cooper-Harper Rating scale, also known as the Handling Quality Rating scale, assesses the ease and precision with which the pilot/aircraft system can perform a well-defined task (Ref. 10). These scales are shown in Figure 2 and 3.

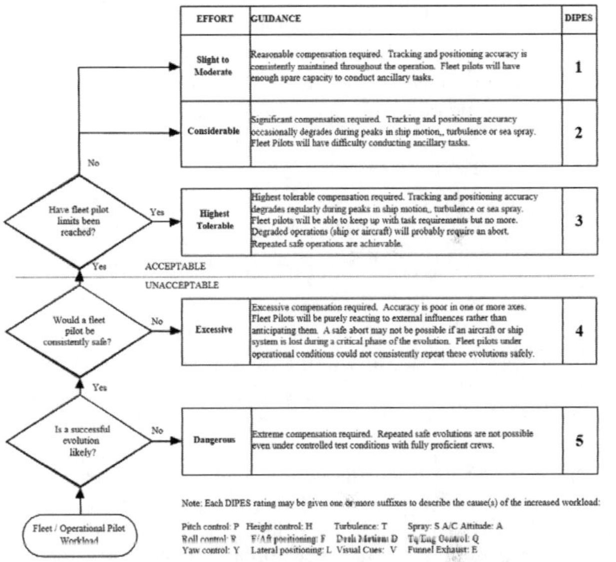

Figure 2. Deck Interface Pilot Effort Scale.

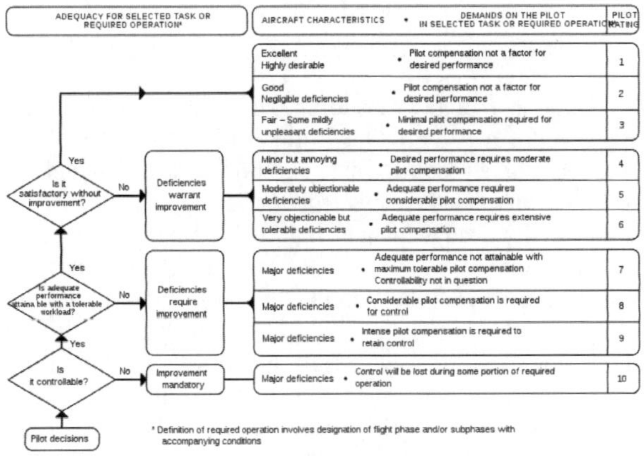

Figure 3. Cooper-Harper Handling Qualities Rating Scale.

HQRs are not typically used during US Navy DI events for rotary wing aircraft, and, as such, established task criteria do not exist. For this exercise, the tasks of interest have been selected as:

Approved for public release; distribution is unlimited, as submitted under NAVAIR Public Release Authorization 2023-50

1. Hover: Maintaining a hover over the landing spot, on the intended landing heading, until a safe landing can be achieved.
2. Landing: Touching down as close as possible to the center of the deck markings.

Tolerance values were selected to indicate whether the task was performed adequately or to the desired standard and are shown in Table 1. The tolerances for the hover represent a cube of space 15 feet above the landing spot within which the aircraft nose gear should remain. The tolerances for the landing are measured via the distances of the nose gear from the intended touchdown point. Success at achieving desired or adequate performance was adjudicated by the pilots with inputs from their respective crew chiefs. The pilot was not given any direct information about position relative to the tolerances while performing the task; they simply tried to maintain the hover as precisely as possible with the aid of crew chief guidance and the existing visual cues.

Table 1. Task Performance Tolerances.

Task	Desired Performance	Adequate Performance
Maintain a hover above the touchdown markings until a safe landing can be achieved	Nose gear remains within a 6ft x 6ft x 6ft cube centered 15 feet above the touchdown markings	Nose gear remains within a 10ft x 10ft x 10ft cube centered 15 feet above the touchdown markings
Touch down as close as possible to the center of the pavement markings	Nose gear touches within a 6ft x 6ft square centered on the touchdown markings	Nose gear touches within a 10ft x 10ft square centered on the touchdown markings

Two pilots completed runs to Ship 1 (Pilot A and B) and to Ship 2 (Pilot A and C). Only Pilot A tested runs for Ship 3. For each ship, a specific wind condition was chosen and then modified in small increments after each run. These incrementally changed wind conditions were randomized and kept undisclosed from the pilot such that they knew the general direction of the Wind Over Deck (WOD) but never the specific magnitude or azimuth. For the purposes of this study, wind conditions were chosen to meet two criteria:

1. Live DI data in a region where a pilot identified at least one DIPES 3 rating and one DIPES 4 rating relatively close to each other, suggesting an area of the envelope where a possible limitation was required.
2. Pilot comments from the live DI event had suggested that the DIPES ratings collected had been primarily driven by the heavy wind turbulence, and not by other considerations, such as cueing or ship motion.

Choosing points that met these criteria meant that it was reasonably likely that the wind conditions simulated should show a change in ratings from satisfactory (DIPES ≤ 3) to unsatisfactory (DIPES 4+) for fleet capability purposes. Finding this barrier between satisfactory and unsatisfactory is the crux of drawing limitations for fleet LRE development, and as such, it was vitally important that the simulator environment be able to replicate these conditions. The exceptions to the above were Ship 1, Low WOD conditions. These conditions were chosen specifically to explore whether pilot rating repeatability and/or sensitivity were significantly different at the edges of an LRE vs in the center. A diagram showing the WOD ranges for each tested ship is shown in Figure 1 for reference.

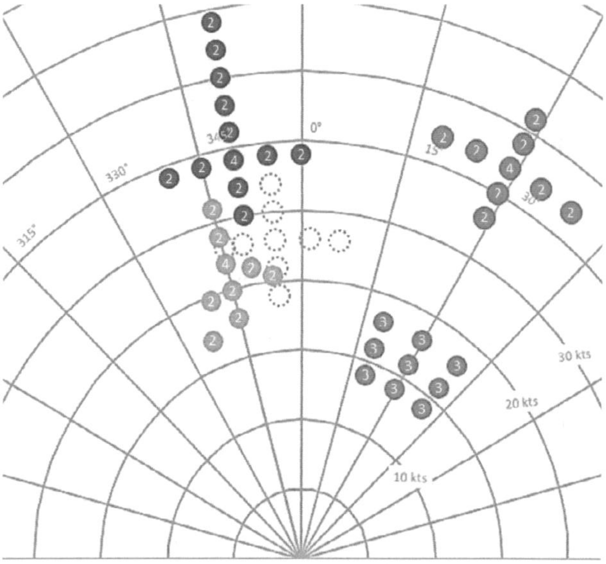

Figure 1. Planned Test Points Wind Rosette.

Figure 1 is broken up into 3 distinct data sets, with red representing data taken at Ship 3, the blue represented Ship 1, and the green showing the Ship 2. The number on each circle shows the number of data points taken at each wind condition, with more data being taken at the center of each cluster than the outskirts for the higher wind conditions. The unfilled in green circles represents the originally planned test points, while the filled in circles are the points actually flown. These points differed from those planned due to anemometer corrections and availability of CFD models. When anemometer corrections are applied to the filled-in circle test points, they will resolve to the unfilled-in circles, which represent what ship operators would observe and utilize.

RESULTS

Over the course of 7 piloted simulator sessions, 191 runs were accomplished, of which 149 were counted as good for score and analysis. 42 test runs are not included for analysis, either because they were practice runs or because the setup was incorrect in some way and had to be repeated.

Virtual Ship 1, Pilot B, High WOD

For the data collected, the team examined how often the pilot would give ratings that were identical, within one, or two or more rating points from each other at identical conditions. This analysis took into account the workload rating given for each unique wind condition, of which there were 9 for this data set, and compared them with other ratings of its kind. For most conditions, this was a comparison between two points, however, some setups had more than one repeat of the condition flown. Table 2 shows the data collected.

Table 2. Results (Ship 1, Pilot B, High Winds).

WOD Azimuth (deg.)	WOD Vel. (kts.)	Hover DIPES	Landing DIPES	Hover HQR	Landing HQR
30	32	2	2	5	5
30	32	1	2	4	3
30	32	1	1	2	3
30	32	1	2	3	4
30	34	3	3	7	5
30	34	2	1	4	3
30	36	2	2	5	7
30	36	2	2	4	3
30	30	1	1.5	3	3
30	30	1	2	3	3
30	28	2	2	4	7
30	28	2	2	6	5
35	32	2	3	4	7
35	32	1	2	3	3
25	32	1	2	4	3
25	32	1	1	2	2
40	32	2	3	6	7
40	32	2	1	5	4
20	32	2	3	4	6
20	32	2	2	4	5

It was quickly noted that intra-pilot variabilities of greater than ±0.5 ratings were somewhat commonplace for these simulation flights. For example, in Table 2, there were four runs at 30 degrees azimuth and 32 kts magnitude. None of the four assessed ratings were consistent across all four runs, with some having variations up to three rating points. To visualize the percentages of these variations, Figure 4 was created. Blue indicates that all of that type of workload rating collected for that wind condition matched. Orange indicates that the overall spread of ratings captured for a wind condition differed by one. While grey indicated these ratings differed by two or more. The number in each wedge shows the number of occurrences. An ideal result from the standpoint of repeatability would be an entirely blue circle, which would indicate that there was no variation in the DIPES ratings. From these initial results that the hypothesis that DIPES ratings would be more repeatable than HQRs hold true; however, it will be shown farther down that this is not a consistent result.

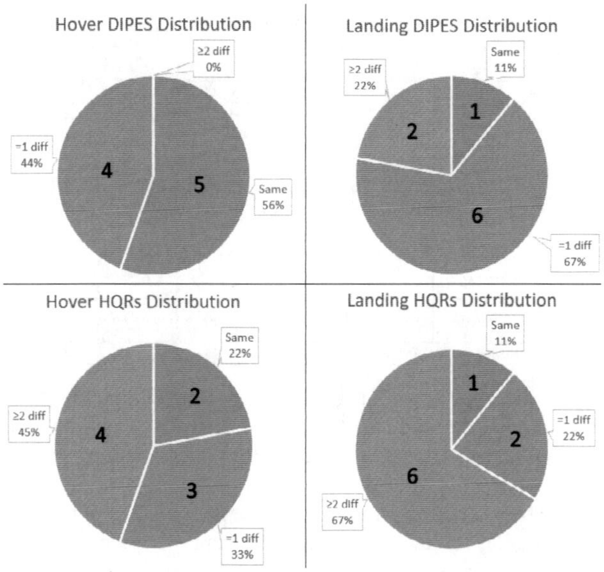

Figure 4. Rating Repeatability (Ship 1, Pilot B, High Winds).

Hover DIPES was consistently within one or less 100% of the time, and landing DIPES being within one or less 78% of the time. These seem to be average for the conditions tested in this study. The HQR consistency percentages were well below expected; HQRs only matched 22% and 11% for Hover and Landing respectively, and ratings were different by 2 or greater 45% and 67% of the time respectively. Table 3 below showed the standard deviations of each rating type followed by an average of those standard deviations across all conditions. There were important elements of this analysis to consider when drawing conclusions from these numbers:

1. These standard deviations were each calculated with a very small data set, usually two numbers.
2. To help compensate for #1 above, an assumption was made that, although the mean rating value at each condition was expected to change, the variance and standard deviation of each condition could be assumed to be the same or similar enough. Averaging ratings across the entire data set would be ill-advised given the differences in mean value across the range of conditions. However, this assumption allowed the team to average the standard deviations themselves across all conditions for each ship / pilot combination to give a more statistically significant number of data points.

Table 3. Rating Standard Deviations (Ship 1) Pilot B, High WOD).

WOD Azimuth (deg.)	WOD Vel. (kts.)	# of Data Points	Standard Deviations			
			Hover DIPES	Landing DIPES	Hover HQR	Landing HQR
30	32	4	0.5	0.5	1.291	0.957
30	34	2	0.707	1.414	2.121	1.414
30	36	2	0	0	0.707	2.828
30	30	2	0	0.354	0	0
30	28	3	0.577	0.577	1	1.155
35	32	2	0.707	0.707	0.707	2.828
25	32	2	0	0.707	1.414	0.707
40	32	2	0	1.414	0.707	2.121
20	32	2	0	0.707	0	0.707
Average Standard Deviations			0.277	0.709	0.883	1.413

Using these calculated average standard deviation values for each rating, the Student's t-distribution (Ref. 11) can be used to calculate the number of points that would be required at each condition to be confident that the sampled mean value would be a certain interval from the assumed "true" mean. As an example, to calculate how many runs would need to be executed to be 90% confident that the mean was within 0.5 (i.e. low enough to identify the workload rating) of the true mean, use the following equation:

$$\text{Equation 1: } CI = t_{\alpha,v}\frac{S_n}{\sqrt{n}}$$

Where CI is the confidence interval, t is the t-value (taken from a table), α is the confidence level, v is degrees of freedom, Sn is sample variance, and n is the sample size. For this effort, the confidence interval was set to 0.5 and the confidence level to 90%. The sample variances are the average standard deviations shown in Table 3. v is defined as n-1, and n is the value being calculated. The result must be solved iteratively. Table 4 below shows the calculations to determine the minimum integer for n where:

$$\text{Equation 2: } \frac{CI}{S_n} - \frac{t_{\alpha,v}}{\sqrt{n}} > 0$$

Which is the lowest number of runs required to have at least a 90% confidence level and a 0.5 confidence interval. Each set of ratings was calculated until the value shown in the table became positive, with the final result of each shown in Table 4.

Table 4. Minimum Runs Required to Reach Desired Confidence (Ship 1, Pilot B, High WOD).

n	t	Equation 2			
		Hover DIPES	Landing DIPES	Hover HQR	Landing HQR
3	2.92	***0.119***	-0.981	-1.12	-1.332
8	1.9	1.135	***0.035***	-0.1	-0.316
11	1.81	1.259	0.159	***0.02***	-0.193
24	1.71	1.455	0.355	0.216	***0.004***

For this given pilot, ship, and WOD combinations, the number of runs needed to achieve confidence in the resultant DIPES rating was 3 for hover DIPES and 8 for landing DIPES. Considering that the DIPES rating can be given at any point in time during the test evolution, the recommended number of runs should be closer to the more conservative value of 8. Given the time requirement to collect this data of roughly 5 minutes per test point, this was within an acceptable and achievable number of test point repetitions for future tests. However, for confidence in the HQR ratings, the standard deviation for Pilot B was much higher. This would require 11 to 24 repeats per data condition; meaning one to two hours for a single wind-spot condition. This number of repeats was not thought to be achievable for a standard simulation study. One may also note that for this task, it appears landing was more widely varied in terms of workload based on both scales. This too was shown to be an inconsistent result for the other data points collected in this study.

Virtual Ship 2, Pilot A

Examining a second test case with Pilot A at Ship 2 provided an example of a different test pilot and a different ship. Table 5 provided the data points and distributions for each of the 9 test conditions, with two to four data points per condition.

Table 5: Results (Ship 2, Pilot A).

WOD Azimuth (deg.)	WOD Vel. (kts.)	Hover DIPES	Landing DIPES	Hover HQR	Landing HQR
345	21.84	3	2	5	4
345	21.84	2	3	5	5
345	21.84	2	1	3	3
345	21.84	2	2	4	5
345	23.84	2	2	4	4
345	23.84	3	2	5	4
345	25.84	4	3	6	5
345	25.84	2	2	4	5
345	19.84	2	2	4	4
345	19.84	2	2	4	4
345	17.84	4	4	6	6
345	17.84	2	2	4	4
350	21.19	2	3	4	5
350	21.19	2	1	4	3
354	20.41	3	1	5	3
354	20.41	2	2	4	4
340	19.64	3	2	5	4
340	19.64	3	2	5	4
337	16.87	3	2	5	4
337	16.87	3	2	5	4
337	16.87	1	1	3	3

From Table 5 itself, it was difficult to determine specific trends or conclusions. As such, a similar analysis was conducted to the previous case, with the following pie charts shown in Figure 5 representing the variation of the workload ratings. Unlike the previous case, which showed that each distinct category of workload ratings was biased to one or two pie-chart wedges, this case shows an almost perfect distribution of variation between the three categories. This result was unanticipated, as it disproves the hypothesis that appeared to be confirmed by the previous section, that HQRs were expected to have a higher degree of variation compared to DIPES due to the larger range of potential ratings in the scale. This data also refutes any correlation previously seen between the landing phase and a higher workload rating spread.

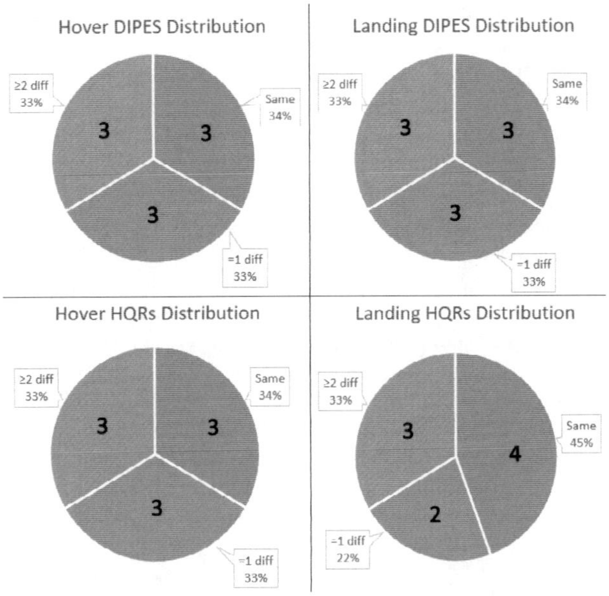

Figure 5. Rating Repeatability (Ship 2, Pilot A).

The standard deviation values were again calculated and are displayed in Table 6. Unlike the previous example, the standard distribution values fell into 3 equal sets, with the standard distribution being either 0, 0.5-0.8, or 1.4 with a few exceptions around the 0.95 range. The average standard distribution values were consistently between 0.55 to 0.7, with the highest variation being in the hover HQRs, and the lowest being in the landing HQRs, which further provided evidence contrary to the expected outcome.

Table 6. Rating Standard Deviations (Ship 2, Pilot A).

WOD Azimuth (deg.)	WOD Vel. (kts.)	# Rated Data Points	Standard Deviations			
			Hover DIPES	Landing DIPES	Hover HQR	Landing HQR
345	21.84	4	0.5	0.816	0.96	0.957
345	23.84	2	0.71	0	0.71	0
345	25.84	2	1.41	0.707	1.41	0
345	19.84	2	0	0	0	0
345	17.84	2	1.41	1.414	1.41	1.414
350	21.19	2	0	1.414	0	1.414
354	20.41	2	0.71	0.707	0.71	0.707
340	19.64	2	0	0	0	0
337	16.87	3	1.16	0.577	1.16	0.577
Average Standard Deviations			**0.66**	**0.626**	**0.71**	**0.563**

The final step for this data set utilized equations 1 and 2 to determine the amount of attempts required to achieve the 90% confidence interval for this test condition. Table 7 shows these values, and as anticipated given the trends seen in this case compared with the last one, the number of required attempts ranged from 6-8 for all 4 ratings, which was a good deal lower than that seen in Ship 1 data with Pilot B which experienced far more variation in the HQRs.

Table 7. Minimum Runs Required to Reach Desired Confidence (Ship 2, Pilot A).

n	t	Equation 2			
		Hover DIPES	Landing DIPES	Hover HQR	Landing HQR
6	2.015	-0.06	-0.024	-0.115	***0.065***
7	1.943	***0.029***	***0.064***	-0.026	0.153
8	1.895	0.093	0.129	***0.038***	0.218

It proved impossible to draw any determinate conclusions from just these two test cases. While the first case showed significantly more variation than the second, both the ship and pilot varied between the test cases, and the total number of repeated attempts required to reach the 90% confidence interval was still higher than anticipated. Each of the remaining data sets showed some similarities but also provided unique results that often did not line up with the two shown here. This may have spoken to pilot experience or background or general ship differences, and while these factors could be worth exploring further, a broader answer to the number of recommended test conditions repeats was desired. Therefore, to better understand the expected workload repeatability, all collected data sets were examined simultaneously.

Summary of Virtual Test Cases

Table 8 below shows the ship and pilot combinations for each of the six data sets, and provides the number of attempts required (n) to reach the desired confidence interval of 90% utilized in the prior mentioned t-distribution analysis. Examining Table 8, there were several comparisons to conduct: pilot to pilot, ship to ship, varied winds with the same ship/pilot, and workload rating to workload rating.

Table 8. Summary of Repeats Required for each Test Case.

Ship	Pilot	Hover DIPES	Landing DIPES	Hover HQR	Landing HQR
		n	n	n	n
Ship 1 High Winds	A	4	5	8	6
Ship 1 High Winds	B	3	8	11	24
Ship 1 Low Winds	B	5	3	13	12
Ship 2	A	7	7	8	6
Ship 2	C	4	4	7	3
Ship 3	A	5	5	9	4

For pilot to pilot comparisons, there were two cases which could be compared, Ship 1 with pilot A to B, and Ship 2 with pilot A to C. For pilot A vs B at Ship 1, it was evident that while the hover DIPES ratings were fairly consistent with 3-5 ratings being needed, the landing DIPES and HQRs spanned much more significant variations. In particular, the HQRs showed 6-8 ratings being needed for pilot A, while pilot B required 11-24 to reach the same confidence level. The Ship 2 comparison between pilot A and C provided a slightly different perspective, with the DIPES ratings varying more with four being needed for pilot C, and seven for pilot A, while the HQRs were much more similar in required test points. This indicated that, in general, different pilots will require different quantities of test points to reach the desired confidence level. It could also point to a difference in how the pilots assign ratings based on the scale. It is interesting to see one rating system providing consistent results, but the other providing a greater spread, denoted by a larger n value. Therefore, the authors believe it is best to conduct testing in the sim with several pilots in order to ensure that sufficient test data density has been achieved without biasing towards one pilot's workload definition. For a future study, it would be ideal to show how the conclusions could change with increased pilot participation and more repeated points to get a more statistically significant sampling of results. The need for multiple pilots to give ratings is by no means a shocking discovery, as it has long been suggested to use multiple pilots for workload studies (Ref. 12), however these results suggest that each condition may need to be repeated more times than previously thought.

For a ship to ship comparison, the same pilot should have flown the same test points at two different ships. However, as seen in Figure 1, wind conditions were varied between ships. Still, pilot A flew to all three ships, and although the wind conditions were varied, a comparison across the three was performed. For the DIPES ratings, Ship 2 had the highest number of test points required to achieve the confidence interval, but overall the number of required runs only spanned 5-7 data points with one exception, which required far fewer repeats. For the HQRs, there was a marked consistency in the hover phase with 8-9 repeats required, while the landing phase only needed 4-6 repeated points. In the case of Ship 1 vs Ship 2, the exact same number of repeats was noted as the desired quantity. Based on this, the evidence suggested that the difference in ship class and wind condition is less likely to influence the rating repeatability than the change in pilots for HQRs, and slightly less likely to influence the outcome of DIPES ratings.

The third comparison desired from Table 8 was to observe the effect of changing wind conditions with the pilot and ship unchanged. For this comparison, pilot B flew two sets of wind conditions at Ship 1, varying from high to low wind magnitudes. Overall, it was very difficult to draw specific conclusions from this comparison, as data gathered by pilot B resulted in far more ratings being necessary in the HQRs than other pilots, but also a large inconsistency in the required DIPES ratings between the hover and landing phases. For the high wind case, the number of data points required went from three to eight between hover and landing, while for the low winds it reduced from five to three respectively. Given the large variation in the workload ratings seen in these two tests, further data would be required to confirm if the change in winds was responsible for the large difference seen in the required test point density.

The final comparison to be drawn was in the workload ratings themselves. HQRs have a larger scale spread, with ratings ranging from 1-10, with most falling between ratings of 2-7. DIPES on the other hand only ranges from 1-5, with ratings of 5 being extremely uncommon in live simulation. As such, there was a significant difference in the span of expected ratings, and as anticipated the HQRs generally required more repeated attempts to reach the desired confidence interval. This is not to say one scale is more accurate than the other, but rather that care should be taken to understand that workload rating scales with a larger range of potential values have a higher likelihood to generate a larger spread and thus increase the number of required test points to ensure rating accuracy. Concurrently, rating scales with smaller ranges will be more likely to fail to capture small changes between ratings that may indicate a workload change.

Table 9. Summary of Hover DIPES Variation.

Ship	Pilot	Hover DIPES		
		Same	1 Different	2+ Different
Ship 1 High Winds	A	5	3	1
Ship 1 High Winds	B	5	4	0
Ship 1 Low Winds	B	2	6	1
Ship 2	A	3	3	3
Ship 2	C	5	3	1
Ship 3	A	5	6	1
Total		**25**	**25**	**7**

Moving on from the analysis of required repetitions, it was also desired to compare the variation of the individual workload ratings between pilots and ships. The most effective method of examining this data is through the pie charts above which showed the number of data clusters where the ratings were the same, separated by one, or separated by two or more ratings. Table 9 through Table 12 delve into this data for each of the 6 test cases by putting the pie chart results in a table format for each workload rating.

Table 9 examines the hover DIPES ratings, and shows that a predominant number of the data clusters were either the same or separated by only one ratings point. The one exception to this was Ship 2 case flown by pilot A, but overall showed that DIPES ratings for the hover phase were likely to span 1-2 ratings. As such, this data indicated that data would likely fall within a range of 1-2 DIPES ratings during the hover phase when tested in the simulation environment.

Table 10. Summary of Landing DIPES Variation.

Ship	Pilot	Landing DIPES		
		Same	1 Different	2+ Different
Ship 1 High Winds	A	3	5	1
Ship 1 High Winds	B	1	6	2
Ship 1 Low Winds	B	5	4	0
Ship 2	A	3	3	3
Ship 2	C	4	5	0
Ship 3	A	5	6	1
Total		**21**	**29**	**7**

Table 10 showed the DIPES rating variation for the landing phase, and had a slight difference from the hover phase. While the overall number of clusters with a separation of two or more in the ratings remained the same, the number of regions with a difference of one between the ratings became more likely. Given this information, it was theorized that the landing phase will experience a higher variation in the workload ratings than the hover phase when tested in the sim. This may be in part due to the length of time spent in hover vice landing, with landing being much faster and therefore providing less information with which to make a rating, or being a potentially harder task. With respect to the brevity of the task, the landing task, which was much shorter than the hover task, would have been more susceptible to the airwake and ship motion conditions at the time than hover. If the "roll of the dice" resulted in a short window of more difficult (or easy) conditions during the few seconds of the landing task, this could have driven variability into the rating not due specifically to pilot-centric factors.

As a whole, DIPES ratings given in the sim should be expected to span 1-2 ratings, meaning that if a majority of test points were rated between a 1-2, with a single 3-rating among them, the workload was expected to be in the 1-2 range despite the point rated at a three. This becomes challenging, however, when considering a case where a majority of the test points are rated as a 2-3, with a DIPES 4 rating occurring once or twice. Usually, a DIPES 4 indicates a condition is not safe for fleet use, but should the fact that multiple DIPES 2s or 3s occurred at the same condition overrule the single DIPES 4 occurrence? This case occurred several times throughout the sim sessions individually, with large data sets being collected. In particular, the case of Pilot A flying to Ship 2 had two instances where DIPES ratings went from a 2 to a 4 at the same conditions between two runs, shown in Table 5. This posed a unique scenario of concern as ratings went from an acceptable to unacceptable level at the same flight test conditions, and further underlined the need for repeated test points at each condition when assigning workload ratings to shipboard tasks in the simulation environment. The test conductor for a given task will need to bear this in mind when determining how many repeats with ratings below a given threshold are required to "erase" a statistically small number of points above said threshold.

Table 11. Summary of Hover HQR Variation.

Ship	Pilot	Hover HQR		
		Same	1 Different	2+ Different
Ship 1 High Winds	A	3	3	3
Ship 1 High Winds	B	2	3	4
Ship 1 Low Winds	B	1	3	5
Ship 2	A	3	3	3
Ship 2	C	2	5	2
Ship 3	A	5	3	4
Total		**16**	**20**	**21**

Table 11 delves into the HQR variations, starting with the hover HQRs. Unlike the DIPES ratings, the HQRs were expected to show a larger percentage of ratings with at least one value difference in the clusters, and the result did show a much larger percentage of ratings with a two or greater spacing between ratings within a group. Given the results of Table 8, it was expected that apart from pilot B, this would be the region in which the largest variation was seen.

Table 12. Summary of Landing HQR Variation.

Ship	Pilot	Landing HQR		
		Same	1 Different	2+ Different
Ship 1 High Winds	A	3	4	2
Ship 1 High Winds	B	1	2	6
Ship 1 Low Winds	B	1	5	3
Ship 2	A	4	2	3
Ship 2	C	6	3	0
Ship 3	A	7	3	2
Total		**22**	**19**	**16**

Table 12 shows the final workload rating, with HQRs taken during the landing phase. Similar to Table 11, the number of clusters with larger variations was higher than those seen in the DIPES ratings, and was comparable to the hover phase, with slightly better repeatability.

Trending Over Time of Virtual Results

Given that the pilots flew similar or identical conditions several times over the course of each event, it was desired to see if their assigned workload ratings were affected in some way due to increasing proficiency with the task. Figure 6 shows the results from this analysis in the form of a pie chart, with the blue section identifying the number of conditions which had the same rating over time, while the grey and orange sections showed a decrease or increase in the rating over time respectively.

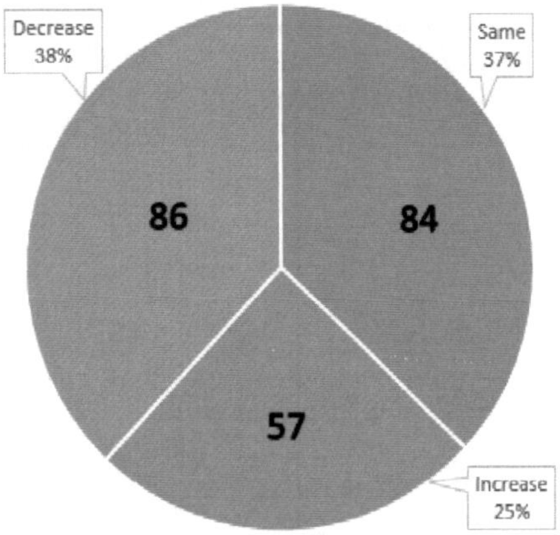

Figure 6. All Ratings Trending Over Time.

Overall, there was only a slight bias towards ratings decreasing (improving) after multiple flights. When all ratings are combined, there was a higher likelihood for ratings to decrease over time rather than increase, but only by 13%. Ratings are just as likely to stay the same with time as they are to decrease. Therefore, there is not much concern that the data in this study was tainted by pilots learning how to perform their task in such a way as to artificially drive down workload.

Live vs. Virtual Simulation Results

While this study focused primarily on the repeatability of workload ratings in the simulation environment, it was desired to compare the results of the virtual simulation to live simulation. While live data was never gathered with this purpose in mind, years of flight test provided a sufficient set of data clusters which were within a small range of azimuth and magnitude which could be analyzed similar to the virtual data. These data points would further require similar gross weights, ship motion, and ideally have as small a gap in years between when the data points were gather minimize control law differences, outer mold line, and approach technique changes. While the gross weight and ship motion conditions compared here were relatively consistent between live and virtual, the live data spanned test pointed collected from 2004 to 2020. A certain level of conservatism should be taken into account when viewing these results due to the effects of the aforementioned influences which may have affected the DIPES ratings.

As with the sim data, individual clusters of live data were examined. Table 13 shows one such case with a cluster of data taken from Ship 3 consisting of 9 test points over a range of 10 knots and 10 degrees azimuth, and virtual data spanning 20 degrees and 14 knots. The gross weight and ship motion values were comparable, and the data was gathered in a single event in both cases, ruling out any differences due to aircraft configuration due to time between test points. While the sim data was collected over a slightly larger range than the live data, it was deemed comparable due to the similarity in the azimuth and magnitude of the wind conditions

Table 13. Ship 3 Data Cluster Comparison.

	Ship	# of Data Points	WOD Azimuth Range (deg.)	WOD Vel. Range (kts.)	Stand. Dev.
Virtual	3	58	340-000	25-39	0.605
Live	3	9	352-001	24-34	0.727

In this example, the standard deviations seen in the data from live and virtual were remarkably similar considering the large difference in test point density. Figure 7 shows the data in a slightly different manner, with the DIPES rating value on the x-axis and the percent of data points within each DIPES rating on the y-axis. This figure demonstrates the distribution of the DIPES ratings as an alternative to the standard deviations.

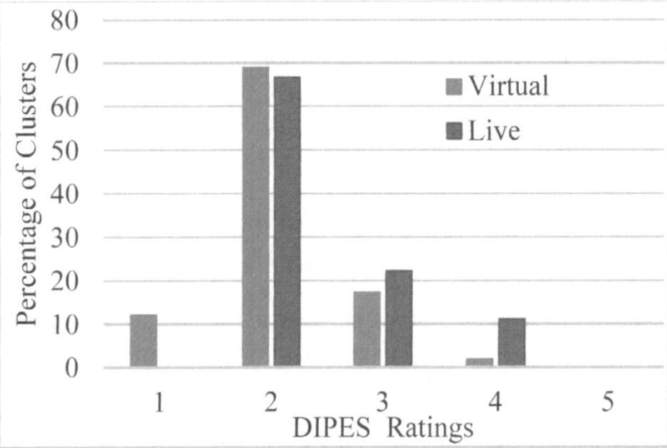

Figure 7. Live vs Virtual Ship 3 Rating Distribution.

From Figure 7, the majority of ratings were given as a DIPES 2 for both the live and virtual test cases. The live data tends to be higher overall, with no DIPES 1 ratings given. However, it should be noted that there are only nine ratings for the live data; there was only one DIPES four and two threes, with the remaining six values rated as twos. The virtual data on the other hand had almost 60 points, with only one rated as a DIPES 4, and the remainder rated as DIPES 1 to 3.

As a second example from a different ship, Figure 8 shows a comparison for Ship 1, with virtual data taken from the high WOD speed condition for both pilots A and B and the live data being from two data clusters. These two clusters were similar to the virtual data, but not considered close enough to each other to combine into a single live data cluster. In this case, the live data appeared to be slightly biased towards lower ratings, while the virtual data had generally higher ratings between a DIPES 2-3. However, the number of ratings from the virtual data was 52, while the live data sets had 8 and 10 respectively. As such, a wider range of values was expected from the virtual data, with one exception. The DIPES 5 seen once in the virtual data as well as several DIPES

4's were significantly higher than the average live DIPES rating. This would indicate that the live DI data is slightly more consistent with ratings in a smaller band, while the virtual data can have a much more significant variation overall. This further emphasized the need for additional test points when conducting virtual events and the desire for additional pilots in order to filter out extraneous ratings that inaccurately depict the test conditions workload.

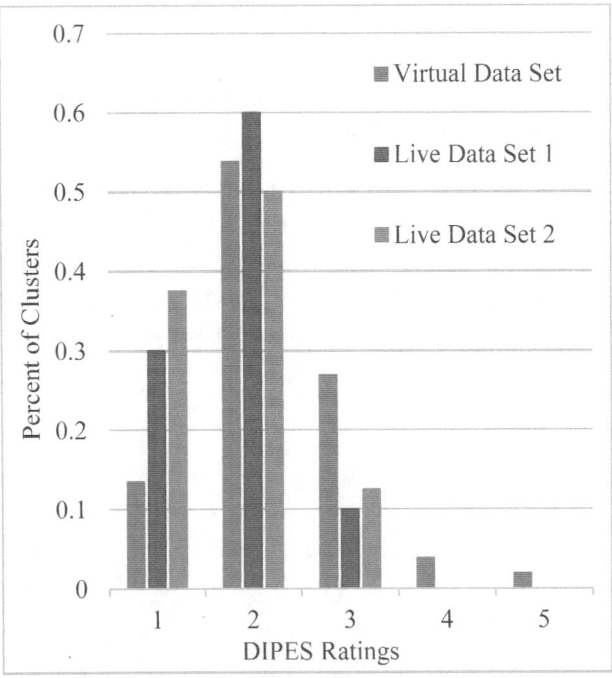

Figure 8. Live vs Virtual Ship 1 Rating Distribution.

The final piece of the live to virtual data comparison was to examine the differences as a whole. Given the available data, not every ship had an equal number of test conditions where groups of data were sufficiently clumped to use in comparison. For the purposes of this report, groupings of data were composed of five or more data points, with wind azimuths and speeds within a small band. Other considerations such as GW range, years the data was collected, and ship motion were analyzed. This yielded no significant trends indicating that low GW and high GW could not be directly compared, or similar comparisons based on age of the data or ship motion.

The team was also able to analyze the live and virtual data holistically; that is, not directly comparing similar setups. The goal of comparing the live and virtual data was to determine if there was more variation overall between the live or virtual cases. The team was able to identify 37 live data sets where the data was grouped close enough together to consider them similar to the groupings used in this virtual study. Figure 9 shows the variation of standard deviations by plotting the percent of times a data set fell within a certain standard deviation band, ranging in increments of 0.1. For the live events, only DIPES ratings were available, as such for each virtual event the data was plotted as red and green bars, showing the hover and landing DIPES standard deviations respectively.

Several items are apparent from examining the distributions. The virtual DIPES ratings were all within 0.2-0.8 in standard deviation, with a majority being in the 0.3-0.6 range. The live data, however, was more spread out, with most data occurring between 0.4-0.6, but with values seen at 0 or greater than 1 as well. The most grouped region of standard deviation values was consistent between the live and virtual data for the most part, however the lack of virtual events to draw from requires a significant amount of conservatism to be taken when drawing conclusions from this figure. Further data sets would be desired to make a conclusive argument, however based on the currently available data it appears as though the largest percentage of standard deviations between DIPES ratings in small groupings by wind azimuth and speed will share a range of 0.3-0.7. This would indicate that a similar number of repeats may be required for live and virtual testing to ensure the equivalent pilot workload was fully identified.

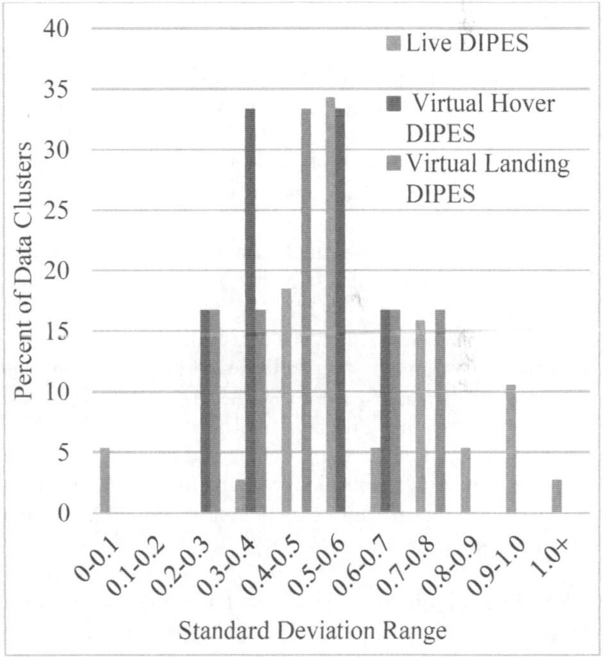

Figure 9. Standard Deviation Ranges of Data Clusters.

CONCLUSIONS

The biggest takeaway from the data collected is that stakeholders should be cautious about collecting one or two data points at a single condition and assuming that the pilot workload ratings gathered from those runs are fully representative of the level of pilot workload required for said condition, as the level of variability has been shown to be quite high. The data also suggests that the pilots' inconsistency in workload ratings tends to overshadow any attempt to pick out a pilot's capability of detecting small shifts in wind magnitude or direction. If a pilot cannot consistently give similar ratings for workload at the same condition, it is expected that stakeholders would need a much larger data set in order to draw any conclusions about that pilot's ability to detect changes in wind condition. Without that, the variability of the data can easily shift any trending and result in data with low confidence of repeatability. Some of this variability might be attributed to the time-varying factor of the airwake model.

For future DI piloted simulator efforts, it is recommended that test plan matrices allow for 8-9 runs at each condition of concern when collecting DIPES ratings. This can most easily be accomplished by collecting 4 runs at each condition with each of 2 pilots, or 3 runs at each condition with each of 3 pilots. For sessions where HQRs will be collected, that recommendation increases to 12 runs per condition. Using multiple pilots is also recommended, ideally 3 or more. These recommendations attempt to ensure that each condition will meet the confidence interval of 0.5 and the confidence level of 90%. These confidence values can be altered as needed by the particular goals of a future study, and the required run numbers can be recalculated using the same methods as this report.

Finally, the team can conclude that, while high pilot rating variability may seem concerning when evaluating the simulator environment, it should be noted that live DI pilot ratings also show a high level of variability. The evidence from the collected data indicates that the sim data may have a slightly higher variance than collected live data. This may be due to shortcomings of the simulation environment or it could be due to the volume of test points gathered in the virtual events. Further testing would be required to gather a conclusive answer as to the cause of workload variation in the simulator environment, but preliminary evidence suggests this issue can be resolved through the use of additional test points and pilots. One suggestion may be to look at the "blind" aspect of this study. A control group could be introduced that is told the wind values prior to each flight. The team is interested to see if this would return more repeatable workload ratings or not.

The findings of this study imply that future piloted simulator DI studies will need to appropriate significant resources, especially in terms of lab and pilot time, when looking to characterize pilot workload during DI tasks. 12 runs at every condition of interest can be time intensive in the simulator, however executing 12 runs in the simulator is still orders of magnitude cheaper and more efficient than executing 12 runs of live DI.

ACKNOWLEDGEMENTS

The team would like to thank the Naval Air Warfare Center Aircraft Division NISE Office and Office of Naval Research (ONR) for being the primary funders and backers of the Dynamic Interface Virtual Environment program. Furthermore, the team would like to thank PMA-275 and PMS-377 for their funding and support. In addition, a thank you to all those involved in the simulation test; especially the pilots of HX-21 who performed the flight tests (both virtual and live) required for this comparison and the SIM operators and support staff at MFS. Special thanks to Dr. Sandi Beaulieu for guidance and advice regarding statistics analysis and Student's t-distribution.

Author contact:
Jeremy Scroggs jeremy.w.scroggs.civ@us.navy.mil
Jonathan Dronfield jonathan.d.dronfield.civ@us.navy.mil
Zachary Smith zachary.f.smith6.civ@us.navy.mil

REFERENCES

1. Williams, LT Suni, L., and Long, Kurtis R., "Dynamic Interface Flight Test and the Pilot Rating Scale", AHS Forum 53, April 1997.
2. Pritchard, James A., Findlay, David B., "Dynamic Interface Virtual Environment Overview", AIAA Aviation, Virtual, 2021.
3. Riley, D.R. and Wilson, D.J., "More on Cooper-Harper Pilot Rating Variability", AIAA Atmospheric Flight Mechanics Conference, Paper No. 1990-2822, Portland, OR, Aug 20-22, 1990.
4. Wilson, D.J. and Riley, D.R., "Cooper-Harper Pilot Rating Variability", AIAA Atmospheric Flight Mechanics Conference, Paper No. 89-3358, Boston, MA, Aug 14-16, 1989.
5. McDonnell, J.D., "Pilot Rating Techniques for the Estimation and Evaluation of Handling Qualities", AFFDL TR-68-76, Systems Technology Inc., December 1968.
6. Smith, Zachary F., "Baseline for Virtual Dynamic Interface", AIAA Aviation, Virtual, 2021.
7. Gaublomme, Donald P., Fean, Tyler G., Calvillo, Robert V., Smith, Zachary F., "Using Virtual Simulation to Enhance Rotorcraft Shipboard Flight Testing", AIAA Aviation, Virtual, 2021.
8. Polsky, S., Wilkinson, C., Nicholas, J., Ayers, D., Mercado-Perez, J., Davis, T. S., "Development and Application of the SAFEDI Tool for Virtual Dynamic Interface Ship Airwake Analysis", 54th AIAA Sci Tech, San Diego, 2016-1771.
9. Forrest, J. S., Hodge, S. J., Owen, I., & Padfield, G. D. (2008). Towards fully simulated ship-helicopter operating limits: The importance of ship airwake fidelity. American Helicopter Society 64th Annual Forum. Montreal, Canada.
10. Cooper, George E.., Harper, Robert. P. Jr., "The Use of Pilot Rating in the Evaluation of Aircraft Handling Qualities". Advisory Group for Aerospace Research & Development, Report 567. April 1969.
11. Evans, Merran, Hastings, Nicholas, Peacock, Brian, "Statistical Distributions", 3rd Edition, Wiley-Interscience.
12. Key, D.L., Blanken, C.L., Hoh, R.H., Mitchell, D.G., and Aponso, B.L., "Background Information and User's Guide (BIUG) for Handling Qualities Requirements for Military Rotorcraft," AMRDEC Special Report RDMR-AD-16-01, December 2015.

AUTHOR INDEX